MERLYN CARTER
BUSH PILOT

Rob Kesselring

rob@robkesselring.com

Other Books by Rob Kesselring:

Daughter Father Canoe
Coming of age in the sub-arctic
and other stories of Snowdrift River and Nonacho Lake

River Stories
Real Adventures in the North American Wilderness

A Rainbow of Peer Helping Skills

Dedication:
To my niece Karen Kelley
a friend of Merlyn and
a faithful companion of mine
for almost a half-century

MAPS

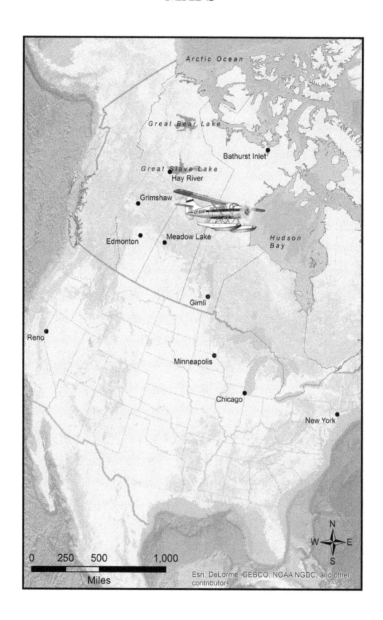

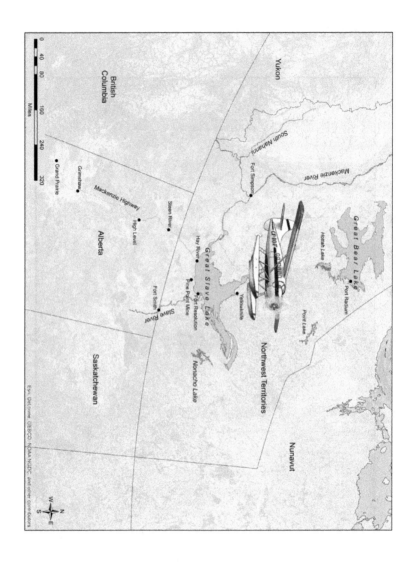

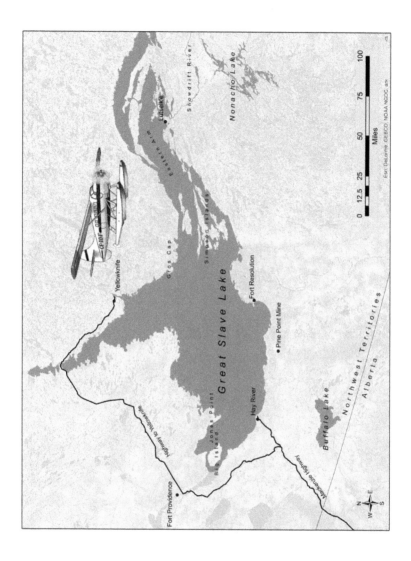

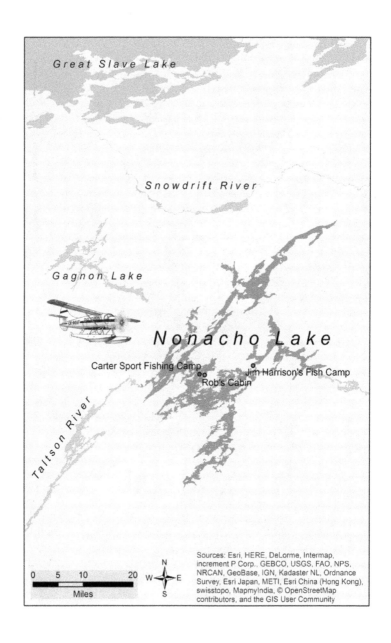

DISCLAIMER

This book details the author's personal experiences with and opinions about Merlyn Carter, his family, and friends. The author obtained information for this book by interviewing family, friends, and employees of Carter Air Service and Carter Fisheries. The author was also a personal friend of Merlyn. A large part of this book is based on that friendship.

The author and publisher are providing this book and its contents on an "as is" basis and make no representations or warranties of any kind with respect to this book or its contents. The author and publisher disclaim all such representations and warranties, including for example any flying stories that may be misconstrued as flying advice. In addition, the author and publisher do not represent or warrant that the information accessible via this book is accurate, complete or current.

The statements made about products and services have not been evaluated by the U.S. government. Please consult with your own legal, accounting, medical, or other licensed professional regarding any choices you make based on the historical stories in this book.

Except as specifically stated in this book, neither the author or publisher, nor any authors, contributors, or other representatives will be liable for damages arising out of or in connection with the use of this book. This is a comprehensive limitation of liability that applies to all damages of any kind, including (without limitation) compensatory; direct, indirect or consequential damages; loss of data, income or profit; loss of or damage to property and claims of third parties.

This book provides a good read for people interested in an extraordinary time in the Canadian North and the exploits of Merlyn Carter, an extraordinary man of uncommon kindness, a true bush pilot. The author struggled for five years to assemble an accurate account of this man's life and times. The author regrets any omissions, inaccuracies, wrong dates, or misquotes.

Reading this book implies your acceptance of this disclaimer.

Table of Contents

Introduction

To understand the North you must suspend your disbelief. You will need to question what you believe are rules of nature, rules of physics and rules of civilization. The North was not opened by following any of those rules. It was not opened by the egg heads, the faint hearted or the couch potatoes. Whether they were white or aboriginal, most of the pioneering pilots, truckers, fishermen, prospectors and merchants had not gone very far in school. They had never read the rules. Merlyn Carter did not know what was impossible. He just did what needed to be done. At a pilots' meeting, a year before Merlyn died, when the topic of discussion was the danger of carrying external loads on bush planes, Merlyn showed a slide of his Twin Otter with two 16-foot Lund fishing boats strapped to the struts flying proudly over the Hay River in Canada's Northwest Territories. A blatant violation of the law, and what might have been thought an unthinkable stunt by southern pilots, earned a standing ovation by the old bush pilots in the crowd. Pioneering northern pilots danced to the beat of a different drum. They worked outside the norms of modern aircraft operation and maintenance protocols, and often outside of the rules and regulations. They just did what needed to be done, usually unaware of the swagger that comes with doing what others believe is impossible.

Modern pilots, even modern bush pilots, might shake their heads in disbelief, but when I say Merlyn Carter once transported thirteen 55-gallon drums of gasoline in a Series 100 Twin Otter, I am not exaggerating. That is what he did. Today's pilots balk at half that load.

Non-fiction is always more unbelievable than fiction because fiction has to stick with what is possible. A biography of a man like Merlyn is not bound by what is possible. I just write the truth. Please don't accuse me of exaggeration. I eschewed exaggeration. This is the real deal.

Sometimes I did need to fill in mundane details. For example, it was sometimes uncertain what exact airplane Merlyn was flying in a story, or what brand of car he was driving when he was pulled over for speeding and whether that happened in Idaho or Oregon. Sometimes the exact year of certain events or who was who, also got confusing. I worked diligently to get the details correct, but sometimes for the flow of the story, I had to fill in what was likely or might be logically assumed. Ballpoint pens were seldom used in the North. In many Hay River homes the only book in the house was a telephone book, and that book was thin. The history of the North is foremost an oral history. When the temperature is 40 below zero, and the wind is howling, and the sky is dark at four in the afternoon, there is plenty of time for storytelling.

Airplane specifications are listed for added interest at the end of each chapter where the airplanes were introduced. This data was mostly taken off Wikipedia. Because of changes in individual airplane's configurations, evolution and modifications, the data may not correspond exactly to Merlyn's specific aircraft and is meant only as a general and helpful guide.

Some data and dates were also taken from Wikipedia to clarify details from oft repeated oral stories especially the Hartwell/Kootook saga.

Airplanes in this book are characters in the story. They had human characteristics--touchy, dependable, fickle. They are referred to by their registrations, which in Canada is five letters. Until more recently, the first call letters were always CF, so nobody bothered with those two. Just the last three letters mattered, sometimes spoken Charlie-Zulu-Papa, but more often just CZP. In this book, for authenticity, I used the same language.

Ninety-five percent of this book is based on eye-witness accounts, many of them mine. Sometimes, I had to choose between conflicting stories of the same event. If I erred, I did so on the conservative side. In a few cases, I have quoted sources directly. I cannot vouch for the absolute veracity of these stories, but if I chose to include them I only did so after being reasonably certain they are true.

This book is foremost a story of a great man. The surrounding factual material included is as accurate as my memory, and the memory of Merlyn's friends and family can attest. I have shared factual information without footnoting or always revealing exact sources partially because it is impossible to differentiate the sources. Information came to me as natural and overwhelming as autumn leaves falling from trees. Many facts are a consensus of years of conversations and experiences. For Northerners so many numbers and dates have so often been repeated and thrown around that they have become part of the public domain. Caribou numbers, mining data, fishing harvests as startling as they may seem to a southern reader are accepted as the matter of facts by the people that live in the North. Readers are free to question what I have written and ferret out statistics in reference books or the Internet. That quest will, for the most part, confirm what I have written. If there is a discrepancy, my first inclination would be for readers to check the veracity of their Internet source. When First Nation people tell me break-up is coming earlier than they ever remember or earlier than what their parents, grandparents, and great-grandparents can remember, or what was passed along in story from even earlier generations, than I don't need a southern scientist to confirm or deny the peoples' observations or their elders' stories. I believe the local people. I confidently share their intelligence with you.

Twenty years ago, I was doing research on paddling the lower Snowdrift River. Looking at maps the drops seemed too extreme for paddlers to navigate with fully loaded tripping canoes. Yet according to a trip report published on the Internet, a group did paddle the lower Snowdrift and without a hitch. The Chipewyan Dene village of Lutselk'e is located on Great Slave Lake near the mouth of the Snowdrift River. When I asked George Marlow, a lifelong resident of Lutselk'e, if he thought the Internet report was accurate, Marlow responded, "I think maybe no." Later I talked with a pilot from a Yellowknife air charter company, he remembered picking that group up in a float equipped Twin Otter from a lake several miles upstream of the river's terminus and above all the dangerous rapids. Is it possible to paddle that lower stretch of the Snowdrift? Expert canoeists and kayakers are doing amazing descents these days including a plunge over the hundred-foot Alexander Falls just south of the town of Hay River. But, was the Internet trip report by those early Snowdrift River paddlers accurate? I think maybe no.

This book is foremost a story of a great man. The factual material included is accurate. If it seems beyond belief, I have succeeded in telling an accurate story. The North is beyond belief, and that is why this story needs to be told. Northerners need affirmation that their history, however unbelievable, is real.

If you don't want to believe what you read, that is your choice. During the 1960s Pine Point staking rush, the breakfast line at the infamous Hay River Hotel "the Zoo" extended far down the dirt street. In that frantic scene, just 200 yards from Merlyn's home. Over a hundred and fifty men were lined up for chow. Another bush pilot legend, Perry Linton, remembers one prospector complaining that he ordered his eggs "easy over," but they came sunny side up. The harried waitress, picked up the plate and dumped the eggs on the top of the unsatisfied customer's head and shouted, "Is that easy over enough for you?"

Merlyn Carter lived at a time and in a place that was as different than the south than can almost be imagined. If you cannot suspend your belief and accept that, well then close this book and empty your breakfast plate over your head. I will not be able to help you understand.

Curl up with this book. Imagine the popping of burning spruce in a tin stove, frost a half-inch thick on the inside of the window panes and a silent sub-zero stillness enveloping the house. Open the book, open your mind, open your heart, and read. I can only hope I did a good enough job with my writing to transport you to the real North and to acquaint you with its real people and its real history. People made the North unreal, beyond belief. Merlyn Carter was one of them.

Merlyn
(photo courtesy of Carter family album)

Chapter 1
Bad Fuel

"Mum, hang on, we're going to crash." These are not the words a mother wants to hear from her pilot son. The plane was wobbling and just 50 feet above the ground. Trees were dead ahead. Two sets of power-lines lurked just beyond the trees. The plane's stall horn was blaring. Its propellor was slowly, impotently spinning. In a clarion voice, not from inside the plane, but from a voice inside his head, Myles could hear his deceased father, Merlyn Carter, calmly saying, "Don't let the plane do what it wants to do, make it do what you want it to. Make it."

So why were Myles, his mother and Terry Webb in this life-threatening aviation pickle? Like most calamities, it was the result of a combination of several unfortunate factors.

Several months before, Webb had purchased his dream airplane a Cessna 206 with call letters VCM. It was a few years old, but what pilot's call low time. Just as cars are measured by miles, the age of planes is measured by airframe hours. Webb's 206 was mounted on amphibious floats. With this plane Webb could land on any one of thousands of lakes and rivers. He could also avoid the logistics of mooring his plane at a seaplane base. With the wheels extended the 206 could land and refuel at airports as easily as a plane with conventional landing gear. Amphibious planes are less desirable for a commercial operator because the added weight of the wheels, tucked inside the pontoons, reduces the useful load. But for a private guy, like Webb, it's hard to imagine a more versatile, safe, or fun airplane.

Webb was still learning to fly VCM and accumulating dual hours to satisfy a requirement of his insurance carrier. He was meeting this requirement by combining pleasure flights and errands, whenever possible, with the commercially licensed, Myles.

On this day, Myles and Webb were on the last leg of a long journey. They had stopped in Hay River to refuel and pick up Myles's mother, Jean, before heading to Myles's son's wedding in Grand Prairie, Alberta, 450 miles south of Hay River. Webb quickly refueled the plane before dashing off for a bite to eat at the Carter's old family home on Vale Island.

The Merlyn Carter Memorial Airport at Hay River has two large fuel tanks which are located just off the apron of the main runway. One is full of jet fuel suitable for jets and turbine aircraft. The other contains 100 octane avgas that works well in most piston engine planes. These two fuels look different, smell different, feel different and are labelled differently. Because of the dangerous consequences of putting jet fuel in a piston engine, many years ago Transport Canada took the precaution of mandating that jet fuel be dispensed through a nozzle that would not fit in a piston powered airplane's fuel tank, in much the same way oxygen and propane tanks have different connections so they can never be inadvertently hooked up mistakenly.

That morning, an old turbo helicopter had refueled in Hay River. To fit the fuel tank on the old helicopter a smaller, outdated nozzle was needed. The new Transport Canada mandated jet fuel nozzle was not replaced after the turbo helicopter was fueled. When VCM pulled up to the tanks, the attendant mistakenly handed Webb the wrong hose. In the rush, Webb did not see the label, smell or feel the jet fuel or check the receipt. So as the three travelers ate lunch, the piston plane sat on the tarmac, filled to the brim with jet fuel, an accident waiting to happen.

Webb had been flying left seat as pilot-in-command, but when the threesome arrived back at the airport and while they were loading up, Jean asked if Myles could fly the plane to Grand Prairie. Webb agreed to be the co-pilot this time, and Myles climbed into the left seat. With full tanks, three people, luggage and the extra weight of amphibious floats, VCM was at gross weight. Still warm from the previous flight, Myles did not run-up the airplane (rev up the airplane's engine while brakes are applied to check its functions) and instead he did most of the pre-flight checks while he was taxiing.

The light winds favored the cross runway 04, a shorter 4500 foot, half gravel, strip that faced the Hay River. Unbeknownst to those in the cabin of VCM, as the plane was taxing, jet fuel was slowly working its way into the engine's fuel lines. Cleared to take-off, the gauges looked good and the plane initially responded powerfully to the throttle. The first inclination that there might be a problem was a slight rattle. It sounded as if the lids on the floats were unlocked, but just as the airplane left the ground Myles looked down at the pontoons and everything looked good, except the rattling was getting louder. Suddenly the fully loaded plane was becoming sluggish, RPMs were dropping, and the plane had ceased to climb. Myles knew he was losing power, he was going to crash. He turned and warned his mother. Had Myles recognized the engine failure a few seconds earlier, he could have chopped power and dropped back down, landing at the end of the runway and beyond. But it was too late for that. He thought of dropping the float plane down on an old flooded channel of the river which ran parallel to 04 and was used when the Carters switched the landing gear on their bush planes from wheel/skis to floats. The wheels on amphibious floats are slow to retract, and Myles knew landing on water too soon with the wheels still extended would flip the airplane. Myles also knew that just on the other side of a grove of trees, across the road, and over the power lines was the present channel of the Hay River where the Coast Guard docked their cutter. If he could clear the trees and the wires, he could set the plane down on the river. The propellor was doing little more than windmilling, maybe 15% thrust at best. Without power, he could not climb. The stall warning horn was blaring, Myles needed some speed to maneuver, but he could not risk pushing the stick forward; any loss of altitude would put him in the trees. Myles could hear his father's voice, "Don't let the plane do what it wants to do, make the plane do what you want it to do. Make it."

Myles skimmed through the tops of the trees, pushed the stick forward just enough to get a little speed and then yanked it back. Although the stick primarily controls the speed of a plane, yanking it back will give the plane a little pop of lift. The plane went up over the wires like an Olympic pole vaulter whose track shoes just sneak over the bar. Long time Hay River fisherman, John Pope, claims it was close, "It was almost as if the floats passed through the wires."

Then Myles pushed the stick forward, gaining speed, avoiding a stall and decreasing the angle of descent. The airplane passed just a few feet above the deck of the Coast Guard ship, with sailors diving off the ship afraid for their lives, and the plane bounced once on the water as it slid into the "dog-leg" channel. It was a successful dead-stick landing on a tiny patch of water.

Mistakes were made which set up this forced landing, but success is seldom measured by the event and rather by the response to the event. Myles had the skills to get safely out of a jam, and he had help, the spirit of his father coaching him in.

"When you are flying an airplane and you want it to do something, and you're in trouble, you make it do what you want it to do. Otherwise, it will do what it wants to do. So if you want it to go to the right, make it go to the right. If you want it to go to the left make it go to the left. Make it. And treat your airplane good. Because if you treat it good, that day will come when you need your plane to treat you good, and it will be there for you."

-Merlyn Carter

Cessna 206 Amphibian Specifications
Data from Wikipedia, AOPA and Wipaire Inc.
General characteristics
Crew: one
Capacity: five passengers
Length: 28 ft 3 in (8.61 m)
Wingspan: 36 ft 0 in (10.97 m)
Wing area: 175.5 ft² (16.30 m²)
Empty weight: 2,708 lb (1,228 kg)
Max. takeoff weight: 3,792 lb (1,720 kg)
Powerplant: 300 hp (224 kW)
Performance
Maximum speed: 150 mph (151 knots) at sea level
Cruise speed: 163 mph (142 knots) at 6,200 ft (1,890 m)
Range: 840 miles
Service ceiling: 15,700 ft (4,785 m)
Take off run (land) 1,146 ft (349 m)
Take off over 50 ft obstacle (land) 1,830 ft (558 m)
Take off run (water) 1,770 ft (539 m)
Take off over 50 ft obstacle (water) 2,850 ft (869 m)
Rate of climb (per/min) 770 ft (235 m)

Chapter 2
You Don't Know Nuthin

I plunk myself down on the river bank. My legs below my knees are still tingling and fuzzy from kneeling. My bow paddler, Peter Lenmark, pulls the canoe up on shore, just beyond the influence of the lazy current. We are close to the big lake now. Great Slave Lake, I can smell it. I can't describe that smell, but it is a smell I recognize instantly whenever I smell it again. I like it. Lenmark and I have come into Hay River by canoe. We left the car at Paradise Gardens just 18 miles upstream. I want Lenmark to experience the town of Hay River first from the water.

By canoe, these days it seems that Hay River is more populous than ever. For eight miles, we pass houses on river left, and just a few miles back, we see the old gray 15 story high-rise towering phallus-like from the town center. That is behind us now. Beached on Vale Island we are amidst rundown buildings and industrial rubble.

"I wish you could have been here in the 1970s Pete." I say it with whimsy in my mind. "There is no way to tell you actually how it was. No way I can make you feel it."

Lenmark is too busy swinging at swarming mosquitoes to pay much attention, and then a big bulldog fly, half the size of a prune, joins the assault. I silence its ripping buzz when I catch it mid-flight. I still have the touch. Squeezing it in my fist I can hear and feel the squish. It isn't like the pop of the beetle or the silent collapse of a mosquito. The death of a bulldog fly is a liquid crunch. I enjoy it, despite the mess in my hand.

"All of this Pete," spoken as I shake the bug's guts off my hand while panning it across the entire waterfront, "was alive with men."

I pause, as I remember, and then go on and speak with enthusiasm, "Bright flashes of arc welders were like fireflies and the ozone? You could taste it. Its sharp odor blended with diesel fumes and dust and permeated the air. The landscape was as alive and chaotic as an ant colony kicked by the toe of a boot. Scurrying men, the staccato beeping of fork-lift trucks whining in reverse, hammering, loads dumping, the screech of steel dragged on steel, foremen barking orders, the hiss of steam. And can you see that dock down there? That was the float base for Carter Air Service. Their de Havilland Single Otter airplanes were so loaded, so heavy, that when they taxied on the river, the floats plowed through the water like torpedoes. When that nine-cylinder rotary engine roared, the water turned to foam. Boaters beware, Merlyn Carter was taking off."

"You saw that big high-rise when we paddled in. It was built in the 1970s and was packed over capacity with people before it was even finished. Some slept in beds, others on foam pads, and some nights the apartment floors were strewn with bodies. Migrants from southern latitudes unable to sleep in the midnight sun would press aluminum foil on the windows. On sunny days the high rise flashed and glittered. A big gas pipeline was going down the Mackenzie River. It was not a question of whether it was going to be built, it was just a question of when. Hay River was the hub of all that transportation. The gravel Mackenzie Highway, the rail line south, the big river northward... Hay River was ground zero. Northern Transportation Company Limited was already busy pushing barges down the Mackenzie and west to Alaska's north slope oil fields. It was also gearing up for the Canadian gas pipeline. Their whole outfit, the only shipping company in the North, was based right here. See the giant building down there a quarter mile; I think it houses the syncrolift. It is only the dry dock in the entire watershed."

Lenmark's eyes followed my gaze. Distracted by mosquitoes uninspired by the flat, drab, dusty terrain, Pete, like many people new to the North, failed to understand my enthusiasm. Had I not been so quick to excitedly chatter, he might have said, "Okay, so we have reached the end of the world, let's turn around and go the other way." But I was already in another world. The spirit of the North had grabbed me again and I blithely chattered on.

"Hay River was the center of all this transportation, of all this preparation. And its reputation of "white collar pay for blue collar work" drew men from all over the world. And most of them came not to save money, but to earn it, and spend it. Maybe you could see it there behind us. I'm not sure, but I think that long building made of squared-off logs is the back of the infamous Hay River Hotel and a bar called The Zoo. I don't mean its nickname was "the zoo." I mean it was really called The Zoo. And it lived up to its namesake. It was the wildest bar in the Territories."

"Women came here too. Tough, bra-less, independent women. They mucked out barges, worked as flaggers, even welders sometimes. They worked as hard as the men. Sinewy arms, faces tanned bronze, they were all beautiful too. Outnumbered three to one by the men and they knew it. After a night of drinking at The Zoo, you didn't take one of those girls home, they took you home."

"So you're saying they were hot," Lenmark said as he lifted his eyebrows with some prurient interest.

"Hot? Let me tell you. I had a job back then putting insulation into the attics of government houses. One time, I had to drag a big plastic hose into a four-plex apartment building. The hatch into the attic was in a woman's bedroom. I remember a poster thumbtacked to her ceiling. It read, "Double my pleasure come again" and it was posted on the ceiling no less. Yeah, the women were hot and they liked being on top."

Lenmark laughed as I continued. He had become more interested in my story.

"There was a classy bar in Hay River too. It was in the New Town. Called the Back Eddy, it served the best steaks in the North. Owned by a New Zealander, Bill Laws, he would sear the steaks first in a deep frier. As a result, the steaks were juicy and tender. My mouth waters still as I think of that red meat. Classy? I guess that is a matter of degree. Laws would bring in some great entertainers from the British Columbia's west coast and all summer long that bar was packed with bodies, eating, dancing and drinking. In that smoke filled, alcohol-fueled atmosphere, more than once, girlfriends slid beneath the shadows of the tables. Wild times, and the Great Slave Lake beach parties? Whoa."

"Tell me about those parties."

"First, let me tell you about some of the characters in those days. Lionel Gagnier was one. No one knew how, but he had the strength of three men. The northern workers were tough, but there was a code of honor. One night in The Zoo a couple of men, too long out on a seismic line and too horny to be polite, rudely accosted a young woman and crossed the line of decency. Gagnier had seen and heard enough of it. He picked the scoundrels up by the scruff of their necks, one in each hand. The doors of The Zoo flailed open and the men's feet didn't even touch the ground until their butts landed in the dust. Oh yeah, there was no pavement in those days. When it was dry, it was dusty, and when it rained, mucky. In fact, from Edmonton to the arctic coast the nickname for rubber boots was "Hay River oxfords." Gagnier? He could open wine bottles by pushing the cork into the bottle with his thumb. He was one tough man."

"But there were odd balls too. There was this one guy, who used to wander around the new town, wearing a trench coat and carrying a shopping bag. Nobody knew what he did to survive or even where he slept, but he was an accepted member of the community. I guess he had an advanced degree in physics from a Toronto University, so we just called him "the professor." Down south he might have seemed creepy but up here he was an accepted member of the community. Beach parties? A sixteen-year-old kid might pass a joint to a 30-year old and it wasn't peculiar to see a high school student making out with their teacher. People would talk and question some behaviors, but not too much."

"A couple miles separates the new town, built after the 1963 flood, from the old town on Vale Island. The transportation industry remained on the island where many people still lived, but all the new housing and that high-rise, that's all up in the new town. Even at forty below zero, people often hitchhiked back-and-forth between new town and old town, usually just jumping in the back of a pickup truck - the ubiquitous vehicle of Hay River."

"Money was flowing in from the government too. Back in the sixties, Prime Minister Diefenbaker believed that if Canada was to keep its sovereignty in the North the government had to spend some money on infrastructure. That push of cash was gushing during the 1970s. Hay River's high school was designed by renowned architect Douglas Cardinal and sided with bright purple aluminum. It may have been a triumph of architecture, but functionally it had its drawbacks and required a never-ending influx of government spending to keep it afloat. Must say though, Hay Riverites were never much for lining up at the public trough. That was for Fort Smith and Yellowknife. Most of the people here were averse to it, and viewed the government as an impediment, not a benefactor. Not to say they wouldn't grab some government dole when they had the chance."

"Perhaps the most notorious Hay Riverite of the time was Don Stewart. He was our mayor, and he did a lot for this town, and for himself. He was also a member of the Territorial Legislative Assembly, (an MLA) which in those days was more of an advisory board than a governing board. The politicos in Ottawa still called the shots, but they let the territorial MLA's cut their teeth on some decisions. At one meeting, the MLA's had to decide where the headquarters would be for the Northwest Territories library system. The logical choice was Yellowknife, Fort Smith, or maybe Frobisher Bay. But the MLA's decided to take a vote and let majority rule. Rumor has it that each MLA wrote their choice for the location of the territorial library headquarters on a slip of paper and put it in a bowl. After the slips were all in the bowl, Stewart made a motion to adjourn for lunch and to count the votes after the break. Stewart was slow to put on his Hay River oxfords on that day and loitered in the meeting room alone for a few minutes before joining his fellow legislators for a meal. When the politicians came back after lunch and counted the votes, astonishingly the library headquarters was awarded to Hay River. You can't argue with democracy."

"Tell me more about those hot women?" Lenmark asked.

But before I could answer we are interrupted. He approaches us in a stagger. At first I believe he is drunk. But these were his words: "I had a stroke ten years past, they said I would be in a chair for da rest of my life. Dat I can't walk no more. But dem words, "never" and "can't" dem not da words of a fisherman." He went on to spit out these words, "I told dem nurses to help me up, and I held em like dis" as he reached out his arms, shoulder height, with wrists curled like an old raven's talons, "And I walked across dat room. I been walking ever since. I heard you boys talking about the seventies. You don't know nothin, you should have been here in the fifties."

*Single Otter with external load
(courtesy of Carter family album)*

de Havilland Canada DHC-3 Otter
Specifications (landplane)
on floats specs would be slightly different
Data from Wikipedia
Produced 1951–1967
Number built 466
General characteristics
•Crew: 1
•Capacity: 9 -10 passengers
•Length: 41 ft 10 in (12.80 m)
•Wingspan: 58 ft 0 in (17.69 m)
•Height: 12 ft 7 in (3.83 m)
•Wing area: 375 sq ft (34.84 m²)
•Empty weight: 4,431 lb (2,010 kg)
•Max. takeoff weight: 8,000 lb (3,629 kg)
•Powerplant: Pratt & Whitney 600 hp (448 kW)
Performance
•Maximum speed: 160 mph (139 knots, 257 km/h)
•Cruise speed: 121 mph (105 knots, 195 km/h)
•Stall speed: 58 mph (50 knots, 93 km/h)
•Range: 945 mi (822 nmi, 1,520 km)
•Service ceiling: 18,800 ft (5,730 m)
•Rate of climb: 850 ft/min (4.3 m/s)

Chapter 3
Fishbox Flyer

George Carter, Merlyn's father, had a keen mind. Although his house in Meadow Lake, Saskatchewan, lacked plumbing, although his brothers were more comfortable skinning a muskrat than reading a book and although he lived at the fringe of civilization, he never thought of himself as poor, uncivilized or destined for a life of physical labor. He was always impeccably dressed, his fingernails neatly manicured, his pants creased and spotless. Even when he had nothing, he imagined himself as having everything, and he behaved as if he had everything. He hired men before he knew how he was going to pay them and bought calves to fatten before he knew how to feed them. He walked with his hands in his pockets, inquisitive, confident, austere. Although he literally did not have a pot to piss in, he was the type of man who could walk into a group of businessmen and play the part of the most astute and successful of the lot.

He dabbled a little in trapping, but soon was buying fur. He learned how to set a gill net, but soon hired a crew of fishermen. He owned land outside of Meadow Lake but hired a man to work it. He bid on contracts to haul freight before he bought a truck. Like many successful businessmen he threw lots of ideas at the wall, hoping one would stick.

His business schemes and inquisitive mind often kept George away from his modest home in Meadow Lake. His wife, Mabel, lost two daughters in childbirth, leaving Merlyn, as an only child.

In rural Saskatchewan, families were expanding in size like Australian rabbits. Siblings were constant playmates. But Merlyn spent much of his childhood alone. He learned from observation and had the ability to remember details. Merlyn also liked doing little building projects, and his peers remarked decades later that he would repeat hand tricks or gestures until they were mastered. At a young age, he learned, through its absence, to value camaraderie with his friends and with extended family members. He was careful not to offend, not to alienate.

Maybe because she had lost two daughters, Mabel was very protective of her son and doted on him. As a youngster, few chores were expected of him. Without other kids in the family, Merlyn grew up oblivious to the responsibilities of childcare and possibly lived his entire life without ever changing a diaper or washing a dish. He learned, by example, and endless repetition, that cooking, household cleaning, and childrearing was the work of women. It was not that he had disdain for domestic duties. It was as if those jobs never even existed. He loved his mother dearly and talked far more with her than with his laconic father. But he did not identify with his mother. His was a protective love. Merlyn adored his mother, but she was never his role model. Merlyn always remembered his mother as fragile. He learned at an early age that it was sometimes better not to tell all to her. By the time he was a teenager his level of honesty with her was truthful, but with many omissions. He did not want to disappoint or worry her.

George treated Mabel with respect and with deference regarding domestic obligations. As was the custom of the times, George spoke little to his son. Nevertheless, without the distractions of siblings, Merlyn often listened in on his father's business dealings. He admired his father's willingness to take risks, to dream big, and to seize the moment.

Aviation was a passion that drove Merlyn from an early age. In the 1930s airplanes were still enough of an oddity in Meadow Lake to draw a crowd, but for Merlyn it was different. He stood and watched carefully as airplanes landed and took-off. He noticed every nuance of aileron movement and flap deployment. As a youngster watching the matinee at the Lux Theater he was more interested in the war newsreels of Spitfires and Messerschmitts engaged in the Battle of Britain than in the feature show.

He once built an elaborate "airplane" using wooden fish boxes for a fuselage, scrap wood-ribbed canvas for wings, and baby-carriage wheels for landing gear. He chose his younger cousin Ray as the test pilot. That experiment ended badly. Merlyn rolled the plane, with Ray in it, down and off his family's garage roof.

Merlyn's first airplane ride
(courtesy of Carter family album)

The plane "drilled" into the driveway. It was before the days of concussions, so Ray just shook off the stars. Merlyn was unfazed, but the crash proved a harbinger of the mishaps to come when he pushed the limits of bush aviation.

George was too thrifty to buy an airplane for fun, but he could sense that aviation would play a larger and larger role in Northern commerce. It was a business decision when he bought a Cessna 140 in 1947. His 12-year-old son was spellbound by the purchase and always eager for rides. George lacked the hand-eye coordination of a good pilot. Observers often described his landings with the little tail-dragging Cessna as adventures. Mabel patiently supported her husband's diverse business decisions and purchases. She had unwavering confidence in her husband's business acumen but was less confident in his abilities as a pilot. She avoided flying with him at the controls and only reluctantly permitted the father/son flights.

Unlike his father, Merlyn was born with innate mechanical ability. He liked to figure out how things worked and was quick to fix a broken bicycle or toy. He also had keen eye-hand coordination. Learning to drive a tractor was far easier and more fun than learning to read. Mabel worried when George took Merlyn up in the plane and would have been angry had she known that her husband let the young Merlyn do much of the flying, even take-offs and landings. Father and son kept that flying to themselves.

As years went by, George's business interests pulled him increasingly away from Meadow Lake, and he traveled more than he was home. Despite his absences, he made sure Merlyn had a little money in his pocket, discipline in his spirit, and a car to drive before any of his Meadow Lake friends.

Friendship was important to Merlyn throughout his life. In his eulogy, I called him my best friend but suggested that a dozen others could say the same of him. After interviewing many people who knew him, I know that estimate was too low.

He will always be remembered as a pioneer bush pilot, but those who knew him remember him foremost as a man of uncommon kindness. Astonishingly, his old friends could not recall a single argument and their eyes would tear up when they remembered him.

How did Merlyn accumulate so many close friends? As a boy, Merlyn played hockey, but was not "the athlete." He was witty and told jokes, but was not the "wiseguy." He was quick to solve problems and innovate, but was not "the Poindexter." He liked music and played the guitar, but was not "the moody musician." He was not a visible leader in a cheerleading kind of way. Shy in front of a group, he often still managed to be the center of attention. Life as an only child with a bit more stuff than most of his friends did not seem to go to his head. He was loyal to his friends, polite to adults and sunny in disposition. That made him well liked and respected. He was an even-tempered, up-for-anything kid, with a bit of wanderlust in his sky-blue eyes. None of those characteristics faded with age.

George and Mabel sent Merlyn, their only child, off to boarding school in the eighth grade, not because Merlyn had shown any particular intellectual aptitude, but rather because George wanted what was best for his son. It is what a successful businessman did.

Founded in 1917, Campion College was in Regina, 350 miles southeast of Meadow Lake. In the 1940s it also contained a small prep school that catered to students from rural Saskatchewan, where opportunities for schooling beyond grade school were limited. When the roads were dry, it was an eight-hour drive from Meadow Lake to Regina. In the spring, when the roads turned to gumbo, the towns were a world apart. Meadow Lake and Regina were also a world apart culturally. Regina was a city with throngs of people, modern distractions and temptations. A couple of Merlyn's classmates were from Meadow Lake, and the three of them embraced the excitement of the city, occasionally escaping from the disciplinary web of Campion School to revel in urban diversions. In the classroom, Merlyn found little excitement. He struggled with reading and often lacked focus with his schoolwork.

Back in Meadow Lake, George had crews fishing the lakes north of town. He also owned a few freight trucks. After trips to New York City and Chicago to connect with merchants and restaurants, he began hauling fish. George's professional demeanor had made a positive impression. His ability to demand top-quality fish from his crews and deliver what he promised earned him a solid reputation. But like a hundred other independent fish companies, Carter Fisheries was small time. George wanted more.

By the 1940s the commercial fishing boom times of Manitoba and Saskatchewan were over. Most of the fishermen were too young even to remember the days when nets were so laden with pickerel (walleye), whitefish and trout that it took two men to pull the gill nets up through the ice. It's an old story. A resource first thought to be limitless was over-harvested. Fish buyers looked north. In the Northwest Territories, there were many large pristine lakes, including two huge ones, Great Bear and Great Slave.

George knew that once the all-weather Mackenzie Highway connecting southern markets to Great Slave Lake was completed in 1949, there would be a chance to go from small time to big time.

The new dirt and gravel road was rough, sometimes snow-cloaked in winter and choked with gumbo mud in spring. It traversed almost 400 miles of muskeg, boreal forest, and trackless wilderness, linking the railhead at Grimshaw, Alberta, to an aboriginal village and Anglican mission where the Hay River joins Great Slave Lake.

George seized the opportunity. He had crews, trucks, connections with fish buyers and a determination to succeed. He was also under-financed, which meant he needed to move decisively and with alacrity.

There was another challenge. This distant venture increased his separation not just from Meadow Lake but from Mabel, who had stayed out of her husband's business deals. But as fishing in Hay River began to boom and with their son at boarding school, she became involved. Mabel developed business acumen that strengthened the growing family enterprise. Commercial fishing fever had now infected the entire Carter family.

Merlyn was always drawn to excitement, and the excitement was to the north, not to the south, at least through his teen eyes. He never liked school, and although he had made some good friends at Campion, how could that life compete with what was happening in Hay River? Planes were flying, money was flowing, and the stories of the wild frontier stirred his soul. At the start of his second year at

Campion, he wrote these words on the inside cover of his school dictionary.

"Merlyn Carter, Meadow Lake Sask. I have quit school today. The date is November 9th on thurs. in 1950. I am going to Hay River to start out in life. Merlyn Carter. The time is 5 pm."

Merlyn was 16, and that is what he did.

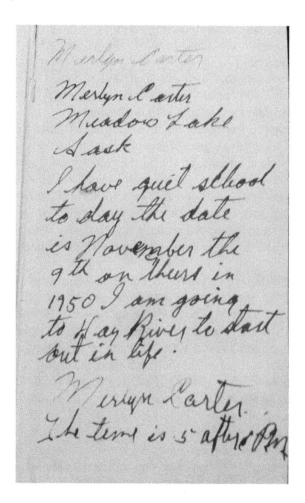

If George and Mabel were disappointed that their son had quit school, they were happy to have one more eager worker, even a 16-year-old boy. These were heady days for the Carters. The fish buying business in Hay River started with a boom and kept booming. The business was expanding like exploding fireworks, but the competition was fierce, and cashflow was often at crisis levels. Lots of money was coming in, but lots of money was going out to acquire boats, trucks, nets, and to build processing plants, warehouses and a dormitory. Despite the abundance of fish, as many fish companies failed as succeeded in those first few years. If George and Mabel were disappointed that their son had quit school, they were also happy to have another set of arms and legs around, even sixteen-year-old arms and legs, to help with the business.

Fishermen on freshwater lakes use gill nets to catch their quarry. The gill net resembles a giant tennis net. It is strung between anchors up to 100 yards apart and left set for a day or two at the most. It is a static net. The net stays put, and fish cruise into it. Think of a strip of flypaper. A gill net counts on fish swimming into the net. The fish pass only part way through the mesh. When a fish struggles to free itself, it becomes entrapped, usually by the gills. The size of the mesh is controlled by regulation so that smaller fish can pass through unimpeded.

On a lake that is free of ice, setting a small single net is as simple as dropping an anchor and an end float, paying out all the net by slowly moving away in a boat. Using the thrust of the motorboat or the push of a canoe paddle the fisherman pulls the net tight and drops another anchor and end float. Small floats on the top of the net and weights on the bottom of the net keep the net vertical in the water. The net can be checked (lifted) by pulling up and along the net with a small boat. After the fish are removed, the net is allowed to sink back into position.

Winter fishing is much more profitable because it is easier to keep the fish from spoiling. Early ice is prime time for commercial fishing, because the fish are concentrated and in predictable locations. It is also easier to set a gill net beneath lake ice that is only a few inches thick. Freeze-up on western Great Slave Lake comes a month later now, but in those days the ice would have been thick enough to fish soon after Remembrance Day, which is Canada's November 11 counterpart to America's Memorial Day. In winter, a challenge fishermen must overcome is stringing a gill net beneath the ice. The indigenous Dene (Athapaskan) people have always set gill nets beneath the ice. Before Europeans arrived, they would manufacture their nets from woven twisted tree bark or thinly cut caribou babiche. They would chisel a big hole in the ice with a bone-tipped spear and then attach a running line to an eight-foot-long pole. Eight feet along, they would chisel another hole, smaller this time, and use a forked stick to push the eight-foot pole another eight feet and so on until the running line stretched the length of the net. Then they could set the net by pulling it through on the end of the running line. Keeping the two end holes open they could keep the set-up going as long as there were fish. Pulling the net and harvesting the fish was similar to pulling laundry off an old pulley system clothes line.

Commercial fishing in the 1940s, in principle, was very similar but the technology had improved. The fishermen used longer nets made from progressively more durable materials. Lead weights and wooden floats kept the nets suspended vertically in the water. The biggest innovation was the jigger. The jigger is an ingenious tool constructed from a long slotted 2x8 plank with a claw and a lever mechanism built into the slot. The jigger would pull the running line along the bottom surface of the ice like an upside-down inchworm. The mechanism to propel the jigger upside-down along the bottom of the ice remains the same to this day and is facilitated by the same measured pull and release of the fisherman.

In early winter, all it took was several whacks with a six-foot steel needle bar to create a hole large enough to angle the jigger in and beneath the ice. An initial careful aiming of the jigger and then the pulling rhythm propelled the running line on a straight course away from the hole. A keen ear can hear the jigger scraping and clawing its way underneath the ice. The jigger was painted red to be more visible through the ice. A second fisherman would follow the sound and depending on the clarity of the ice, and by sweeping away the snow, keep an eye on the jigger. When the operator of the jigger reached a mark on the running line that corresponded to the length of the net, he would stop. The jigger would be positively located and a few whacks of the needle bar would open another hole. A broomstick with a long stout wire ending with a u-shape would snag the running line. The fishermen could then set a net.

The net is lifted by pulling it up through the hole and stretching it out on the ice, being careful not to pull the running line completely down through the opposite hole. In this way, the net could quickly be reset without using the jigger. Running several nets in a long line minimized the amount of chiseling as the same hole could be used to end one net and start another. Fish are removed from the net's webbing with a home-made tool consisting of a small L-shaped claw affixed to a handle, often a piece of an old hockey stick. The fish are thrown on the ice to be gilled and gutted before being loaded into the planes.

Needle bars to fracture the ice and chisel-ended bars used to widen the holes could last forever unless the small loop of cord at the top of the bar, still called by some, "the bitter end" slips through the grasp of the fisherman's hands. Lose the grip and the expensive bar would dart to the bottom of the lake. Hand and power take-off augers eventually replaced some of the labor of punching holes.

Frigid December temperatures would push the ice down another inch every night, until by mid-March when it would be five feet thick. Even at that depth skilled fishermen could still set new nets but chiseling the holes and following the jigger became progressively harder. In the early season, every day counted.

Although back in the 1940s wooden fish boxes were not used in the airplanes, they were used in the trucks, and George had his son nailing them together by the dozen. Soon Merlyn went from nailing fish boxes to driving trucks, from Hay River to the rail head at Grimshaw, Alberta, often in tandem with another driver. Getting fresh fish to market was a race against time and weather. Blizzards could close the road, and a delay might result in the fish freezing. A warm spell could thaw the road and the resulting gumbo could reduce progress to a crawl, risking spoilage. The overload of fish in the truck was almost more valuable than the truck itself, so every effort was made to keep moving.

Great Slave Lake (photo by author)

Other fish companies had trucks, connections with urban fish buyers, established crews, and better financing than Carters. What gave George a leg up? In large part, it was his vision of the value of aviation to the emerging northern fishing industry. The airplane enabled his crews to start fishing earlier, to move crews quickly, and to harvest interior lakes in addition to Great Slave Lake.

Just a year after quitting school, Merlyn earned his private pilot's license and was flying right seat with a legendary bush pilot, Stan McMillan, and hauling fish for his father in a Fairchild 82.

Although George and Mabel kept a house in Meadow Lake, they also had a place in Hay River and spent more and more time in the Northwest Territories. After leaving the Campion School, Merlyn now bounced back and forth between his boyhood home and the North. George was also involved in summer fishing with boats and packers, but most of the fish flying was winter flying. Merlyn would go back to Meadow Lake for much of the summer. He would arrive in his hometown with a splash, and a pocket full of cash.

At age 18, George trusted his son to fly the company's new Anson down to Swift Current for its Annual Certificate of Airworthiness (C of A). That twin-engine warbird was not easy to fly. But because Merlyn had learned from McMillian, George knew he was ready. He often had high expectations of his son. Merlyn met them with little introspection.

An older Merlyn was once asked by his friend Tony Jarrod, "How was it working for your father?"

A bit flummoxed, Merlyn paused before answering. "Of course it was good. When your dad buys you an airplane and supplies you with fish to fly, how could it not be good?"

Almost as much as flying, Merlyn loved music. The life of a pilot is hurry, hurry, hurry, wait, wait, wait. When the plane was loaded, and the weather was good he flew. When the weather socked in or the engine needed a part, Merlyn waited. Other pilots may have brought a paperback or magazine along in the plane to read. Merlyn brought his ukulele because there was not enough room in the cockpit for his guitar. He had an ear for melodies and a disdain for sheet music. He learned to play instruments as he learned to fly, by watching others and by playing and playing and playing. At age 19, he arrived home in Meadow Lake for the summer, flush with $2,500 in his wallet. He borrowed a truck, drove to Saskatoon and bought a piano. He played that piano, and his sons played that piano, for over a half-century.

Merlyn liked the sound of a bottle of beer opening, but seldom drank more than he could handle and never became belligerent when intoxicated. If there was a party, Merlyn found it. At age 20, he convinced three of his pals to drive to Reno, Nevada. In Merlyn's pale green Plymouth Cranbrook Club Coupe with the windows down and pedal to the metal, they drove south in youthful glee, throwing beer bottles out the window, looking for and finding fun. The fun was almost cut short in Oregon.

"Hey, bud do you have a pilot's license?" asked the Oregon state police officer.

"As a matter of fact I do," said the smiling Merlyn as he pulled his private pilot's license from his wallet. That same endearing smile that had convinced his mum that he would be fine on a 2,000-mile road trip in his green sedan, now brought a smile to the trooper's face, and just a warning to slow down. No speeding ticket for Merlyn, and no interruption in the fun.

Riding on a dusty Nevada highway, Merlyn asked his shotgun riding friend, "Who's that girl that's taking the tickets at the Lux Theater? I saw her there last September."

"That girl? That brunette? That broad? You have been away from home too long. She's not taking tickets anymore; now she's the Popcorn Girl."

"You have been gone, she's a farm girl from Golden Ridge out by Goodsoil. She's been around a couple years, but there's been a change in her."

"Whadda ya mean?"

"Damn, you saw her, damn. Why? Look at her. That's Jean Lay? She is the plum of Meadow Lake!"

"Yeah, I saw that much." and a flight plan was filed in the young pilot's mind. "I am going to look her up at the Lux when we get back."

Merlyn in front of Avro Anson
(photo courtesy of Carter family album)

Cessna 140
Data from The Complete Guide to the Single-Engine Cessnas, AOPA Pilot, and Aircraft Specification No. A-768
Introduction 1946
Produced 1946-1951
Number built 7,664
Unit cost $3,495 USD
General characteristics
•Crew: one
•Capacity: one passenger
•Length: 21 ft 6 in (6.55 m)
•Wingspan: 33 ft 4 in (10.16 m)
•Height: 6 ft 3 in (1.91 m)
•Wing area: 159.3 sq ft (14.80 m^2)
•Empty weight: 890 lb (404 kg)
•Gross weight: 1,450 lb (658 kg)
•Fuel capacity: 25 US gallons (95 litres)
•Powerplant:Continental C-85 four cylinder 85 hp
Performance
•Maximum speed: 125 mph (201 km/h; 109 kn)
•Cruise speed: 105 mph (169 km/h; 91 kn)
•Stall speed: 45 mph (72 km/h; 39 kn) flaps down
•Never exceed speed: 140 mph (225 km/h; 122 kn)
•Range: 450 mi (391 nmi; 724 km)
•Service ceiling: 15,500 ft (4,700 m)
•Rate of climb: 680 ft/min (3.5 m/s)
•Wing loading: 9.1 lb/sq ft (44 kg/m^2)

Chapter 4
The Plum of Meadow Lake

The forging process a blacksmith uses to transform a rod of steel into a tool, first involves extreme heat, followed by heavy and focused pounding of the hot steel with a hammer on an anvil. This process alters the molecular structure of the steel. It creates an alignment of molecules and no matter how that tool is filed, finished and buffed the internal strength; the unwillingness to lose its edge and the resistance even to bend is forever part of the character of the completed tool.

Born in 1879 in a Tennessee holler to a clan of meat cutters, Jasper Lay moved westward with his family while he was still a boy. His father was lured by stories of free land, independence and prosperity. He sold the family butcher shop, gathered up his immediate family, some siblings and cousins and set his sights on homesteading in Oklahoma. They imagined that Oklahoma would be similar to the land they passed through in Missouri — green fields, fat cattle feeding in verdant pastures, stands of hardwood. The young Jasper was frustrated when the clan settled in the Oklahoma hill country after discovering the land set aside for homesteading had already been taken. Most of the family resigned itself to subsistence living and what they could earn as farm hands and laborers. Not only that, but the Oklahoma land was harsh, prone to drought, and already overcrowded with homesteaders. The promise of prosperity proved false.

Young Jasper could not swallow that disappointment. Many men endure a gnawing hunger, but fear of the unknown, of leaving a sure bet for a chance at something more, keeps their jaws clenched, unable to take a bigger bite. Jasper had in his blood, the courage to chase his dreams and assuage his hunger. Unlike most of the rest of his family, he refused to resign himself to Oklahoma and to settle for working someone else's fields. He wanted independence and was willing to pursue it with a singular and impatient passion that even exceeded his father's. In 1896, at seventeen, he married a fourteen-

37

year-old bride, Esther, and began his own odyssey — a quest for land and independence that would span 34 years and two countries. In that geographical area at that time, those ages were not outside the norm for matrimony. What was unusual was Jasper's courage to chase his dream, wherever that quest would take him.

The DNA that was to forge his granddaughter's character was fierce determination and perseverance, like hot steel glowing red. Ardor was melded into that family's DNA and it would not cool. Jasper Lay refused to settle for less than what he believed was possible and justly deserved. Soon to be known only as "Vater," Lay eventually led eight sons, two daughters, spouses, some cousins, even grand babies to travel across the continent with little more going for them than kinship bonds. But these were bonds of steel, as strong as aircraft cables. Worldly possessions were few, stashed in a couple of trunks dragged back and forth first across the American west and then the Canadian prairie. Pushed by his dream of independence and land Vater always seemed to arrive too late for homestead lands and too broke to buy land. He still imagined owning land like he had witnessed in Missouri - green fields, fat cattle feeding in verdant pastures, stands of hardwood. He refused to let go of that dream.

For several years the prairies of Canada were a disappointment to Vater. A few years here, a season there, always chasing false rumors. Once he almost secured a purchase of land as far northwest as Peace River, Alberta. Then there was the Fernwood mistake and a missed opportunity in Hubbard. The family, discouraged but not defeated, wandered and wandered, always just missing out. When owning land was not possible, the clan lingered only long enough in one locale to build up cash reserves and perhaps birth another baby or celebrate a marriage, before moving on.

Finally, after over three decades of knocking around in a fruitless search for land and independence, the 51-year old Vater, his sons and daughters, now adults themselves, caught word of new lands opening up for homesteading in a place called northern Saskatchewan. A place where lands were rumored to be vast, well watered, fertile. Saskatchewan, where the Dominion motto was, "a new school built every day," was booming. Settlers from around the world streamed in, flooding the land with optimism and hope. How the Lay family, these Tennessee wanderers, exactly crisscrossed the prairies and established residence in Saskatchewan remains a mystery. But for the last 400 years, and even to this day, if you want to move to Canada, if you really want to, you can find a way. And that is what Vater and his clan of Lays did early in the twentieth century.

Their search was over. The land to the north had just recently been surveyed and in 1930 was open to the Lays for homesteading. Their long quest would be over if they could get there and stake their claim. Each adult could claim 160 acres. Between Mother and Vater, and all the offspring, they arranged to claim as much as three sections of land, almost 2000 acres. Having missed out in Oklahoma, Alberta, Manitoba and southern Saskatchewan, they would not waste time now. In late August with five cows and the family now swollen with more spouses, grandchildren, a pony, a few chests of hand tools and kitchen crockery, they boarded a train northward moving closer to Vater's and his family's lifelong goal, owning their own land. One of the sons, Lloyd Lay, bragged, "I have five dollars in my pocket, a pony, and a dream, and that is enough."

Based on the American Homestead legislation the Canada's Dominion Lands Policy was already sixty years old. It was rooted in the premise that a family could make a good living on 160 acres of virgin prairie. The government was eager and the railways, more so, to promote this idea and lure homesteaders with the promise of almost free land. The aridity of the land west and northwest of the Mississippi was not taken into account, and many homesteaders had blind faith that they could transform the dry plains of Saskatchewan into the fertile valleys of Ohio. The first decades of the twentieth century were a time of frenzied homesteading in Canada. Five thousand new schools did open in Saskatchewan, and land was claimed almost before survey crews could stake it. Despite the seemingly endless horizon of prairie, legally securing a homestead frustrated more potential settlers than it satisfied. The prospect of almost free land drew in droves of people and many were snared by red tape in a haze of the most primitive communication network.

People of many different nationalities, speaking different languages, were invading many landscapes still inhabited by First Nations people. Untouched by the plow and with only the most rudimentary infrastructure it was a wilderness. All of this was happening on the cusp of the great drought, the dust bowl, and the Depression.

It was a ragtag outfit, the Lay clan, that boarded a train in southern Saskatchewan September 1, 1930. It's hard to determine how many Lays came in the first wave, but the legend is that everyone came together. Written record does not exist. In the stops and starts from the departure years ago from the hill country of Oklahoma, spouses joined the ranks and babies were born. But it would be a mistake to believe that this was a loose, aimless or shiftless crew. It was not a democratic family either. Quite the contrary. It was patriarchal. Vater called the shots, kept track of the cash and kept the peace. Fisticuffs may have settled disagreements between brothers, but family loyalty was never questioned. Sharing of resources between one another was so ingrained as never even to be questioned. Babies were as likely to be suckled by an aunt as by their mother.

Established homesteads dotted the broad prairies the Lays passed by on the train northward. Some were built of logs cut and dragged from forested river bottoms. Others were made of sod sliced right from the prairies. The fall harvest was already in full swing, and the summer rains had been plentiful. The crops looked robust and filled the hearts of the Lays with hope. As the train chugged northward toward its destination in St. Walberg, the prairie started to give way to stands of poplar and ridges dotted with jackpine and even some spruce. Farther north the train chugged along. Forest from the North Saskatchewan River valley extended deeper and deeper into the prairie until when they arrived in St. Walberg, it was prairie no longer. The topography was a parkland of forest punctuated with pothole lakes and meadows of river grass. Poplar trees thickly cloaked the vast landscape, and thorn bush grew so tightly that to penetrate the forest on foot looked impossible.

St. Walberg was the end of the rail line, but not the Lay's destination. The land they wanted, the land they had preliminarily signed for, was just recently surveyed and 65 miles farther north. North into what, in 1930, was a trackless wilderness.

September is a seductive month in Saskatchewan; the days early in the month are still longer than anywhere to the south, and the sun seems reluctant to let its summery grip slip away. There is a temptation for residents to relax, breathe deep and relish their apparent warm conquest of the land.

Nature never waits for long. The satisfying smell of the harvest does not dupe the geese. V's of them stream overhead in early September, fleeing southward. As days pass, mornings start to flirt with frost, at first just a hinting blush of white on a few strands of marsh grass at dawn, but later in the month with enough structure to crunch with each footfall. The blanketing fog that was so quick to burn off in July now lingers almost until noon. Newcomers may revel in summer's last gasp, but the aboriginal people listen as evening breezes whistle through the pines, whispering, "winter is coming."

The Lays perhaps did not fully realize they were not in Oklahoma or even southern Saskatchewan anymore. They were too impatient and too headstrong, too determined to be distracted by whispering trees. As daunting as that seemingly impenetrable forest may have appeared, there is no evidence that the coming of winter discouraged the Lays. Only 65 miles separated them from their promised land, and they were quick to charge forward.

This is not a modern family camping story. This journey was a story of survival. Women carried their nursing babies; children stumbled along; cows were coaxed over deadfalls; all the tools needed to clear a homestead and build a home were carried on the backs of the men. In the first few days of their journey some of the forest had been cleared to fuel the insatiable appetite of steam locomotives, but the slag and emerging new growth sometimes made progress even slower. Trails and rutted paths led to recently claimed homesteads, but these routes were often off-course or petered out into game trails of moose or old buffalo wallows. Lacking roads or even trails, navigation was accomplished by compass headings and toil. The Lays were not necessarily able-bodied. These pioneers had a history of hard, dangerous labor and minimal doctoring. All of the sons had nicknames, and one of Jasper's sons who also was named Jasper was usually known only as Barney. Barney's left arm had been severely damaged in a farming accident when he was a boy. A horse had stepped on him, and his arm fused at the elbow and remained rigid the rest of his life. That disability did not stop him from hoisting a trunk on his back and pushing forward. Canvas satchels of tinned goods, awkward loads of boxes and cases, long-handled hoes, pick axes, and shovels, all were packed on man, woman, and beast. In the midst of autumn rains, slickers of waxed cotton were damp and heavy, canvas tarpaulins dripped, cotton clothing clung to shivering skin and leaky boots led to feet blistered with trench rot.

In the evening, maybe the family could find some higher ground to rest and camp, and after felling dead jackpines, they were rewarded with robust and illuminating fires. Just as likely, they made camp in the poplars and sat beside smoldering fires with little heat. Vater told tales of his boyhood woodlots of hickory in his native Tennessee, but they were in Saskatchewan now. Pine fires spat sparks and burned quickly to ash. Green poplar smoldered and hissed. There were no glowing coals to push out the dampness of their clothing and warm their bones. Rainy days led to evening chills with babies and adults alike, slumped up against supine and steaming cows, who shuddered at the chorus of wolves lurking and watching from the thick woods. The sun was down for almost 12 hours, and the only light in the darkness was the campfire or the smudgy glow of a lard oil lamp.

The wilderness was traversed as one foot was placed ahead of the other, and as one day led to another, mile after mile. Reaching the Beaver River led to mixed feelings. It meant they were close to their final destination, but crossing it looked perilous. To ford a river without any clue of where or how must have tested the clan. Many in the party could not swim and wrapped their arms around the necks of the cows or the pony. Although not swift or deep, half-way across, the current tugged at them; the cold water numbed their feet into clubs. It is impossible to underestimate this swirling obstacle full of hidden roots and dangerous undercurrents. Wet, bedraggled, exhausted they successfully crossed the river. This same river would claim the life of Uncle Patch Lay's fiancée just a few years later. Uncle Patch, Vater's brother, would never find a bride but was a favorite of his nieces and nephews.

The seven-day journey with babies, livestock, and tools would alone be a notable odyssey, but that was just the beginning. There was no sanctuary at the site of the homesteads. It took over a week just to find the survey stakes, so hidden they were in the forested sections. The Homestead grids were just arbitrary geographic demarcations. Sandy meadows, thickets of spruce, grassy bottomlands — homesteading in northern Saskatchewan was a sight unseen land grab lottery. To own the land, homesteaders had a to pay a fee of ten dollars per hundred and sixty acres. They also were mandated to build homes, clear and farm the land in a limited time frame. For an extra ten dollars, they could own the mineral rights to the property. Vater declined that expense (a decision descendants would rue almost a century later).

There were much more immediate challenges in 1930 than worrying about mineral rights. Before the winter snows arrived, and that could happen as early as October just days away, meadow hay by the rivers needed to be cut for winter livestock feed. Shelter for the family and the animals needed to be built. Not with lumber from a lumberyard, but these structures needed to be built from the forest, all with axe and crosscut saw. And, the family needed to eat. In this wilderness, it would be a myth to believe that living off the land was possible. The Hudson Bay Company had learned long ago that southern Saskatchewan had the herds of buffalo and far northern Saskatchewan harbored wintering masses of caribou. But in between? Just a few moose and starvation. Nevertheless, the Lays were a resourceful lot. Jasper's marksmanship, honed shooting possums out of the Appalachian treetops, was passed along to his sons and that skill was now employed to kill anything that crawled or flew in the wilderness. Fortunately, 1930 was a peak grouse year, and these wild chickens that follow ten-year cycles of boom and bust were abundant. To save precious ammunition, most of the grouse were snared. In this wilderness, the birds were unfamiliar with humans and could be approached to within a few yards. A snare on the end of a stick was draped over their heads and snagged tight. A plump grouse was a free few mouthfuls of lean protein.

It was November and snow was already accumulating in the shade, when Vater sent two sons to the town of Meadow Lake to register the homesteads and to bring back a 100-pound sack of flour. It was a fifty-mile walk, almost half of which was unmarked. The boys made the round trip in four days, taking turns carrying the 100-pound flour sack and keeping themselves warm during the long November nights by burying their bodies in haystacks piled by other homesteaders along their route. The flour was cherished, but contained nowhere near enough calories to sustain the family that first winter.

A Saskatchewan winter? Weeks of sub-zero cold, howling winds, and stinging blizzards. The boys might have shot a moose. The First Nation people say that in the winter if you jump a Saskatchewan moose from its bed, and if you have snowshoes, you can walk it down and will eventually get the shot. This strategy is not true with caribou, and even with a moose it may take a few days to walk one down, and it may be a long way from the homestead by the time you gut and quarter it. The Lay boys certainly had the tenacity and the backwoods skill to perform this feat. But moose were not abundant in the 1930s. They had been hunted doggedly by Indians, homesteaders, and trappers. Even if the Lays had managed to kill a moose or two that first winter, it does not take long for a big family to eat a moose.

That first winter the family crowded into a rectangular log cabin. It was more a barracks than a home. Built hastily of damp, green, bark-on poplar logs and chinked with moss, the stone fireplace mortared with mud, it could not begin to drive out the moisture. A roof of sod, windows of paper, winter light from the weak, low angled sun and a single oil lamp kept the interior in a dim, smoky haze. A few candles saved for precious Christmas night illumination were the only vestige of civilized accoutrements. It was only the fortitude of the clan and a little music-making that kept the soul of the Lay family alive until the New Year, but fortitude does not fill hungry bellies. There was one quality the Lays possessed that got them through that first winter alive. One thing and it was one thing that was the opposite of everything else that defined the Lays. To a family that had always done it their way, with grit, (without charity) and with a family bond as tight as welded steel, it was one thing, a small act of generosity, that kept them alive through the cold months of 1931.

Although the checkerboard of homesteads created by the Dominion Land Policy looked neat and tidy on a map, by its very design, the orthogonal grid, it isolated neighbors from each other. Even decades later, families lived miles apart and running to a neighbor to borrow an egg or to get help hoisting up a slaughtered hog was not practical. Of the 5000 Saskatchewan schools holding session in 1930, most of them had fewer than a dozen students on their roles. This suited the Lays fine. Their family was large and growing. It was almost a community in itself, and theirs was a family pride bordering on arrogance. They could do anything, and they could do it better than anyone else. It was ironic that in that first winter their survival depended on critical generosity from a neighbor.

A few miles from the Lays' sections, through the woods and across the meadows was a more established homestead. The owners of that homestead were unable to get their entire potato crop harvested before winter closed-in and froze the ground four feet deep. It is unclear how the Lays got the word, or whether the other family reached out, seeing how desperate the Lays were for food. But an offer was made, "There are potatoes, many of them, frozen rock solid in the frozen ground of our fields. They will be mush in the spring, but if you pull them now and drop them, still frozen, into boiling water, they will cook as if they are fresh. You are welcome to take what you need." It was an act of generosity that cost the neighbor nothing, but it saved the Lays. All through February and deep into March, with pick-axe and hands blue from frostbite, the Lays wrestled potatoes from that frozen field and survived.

Rob Kesselring

To know hunger, to really know hunger is something never to be forgotten. Never again, even in the depths of the "Dirty Thirties" did the Lays ever need help to feed the family, but they did that one time, and they remembered.

Today, looking at the sections which the Lays' homesteaded in the 1930s, it seems almost impossible that it began with an axe, a hoe, and a few strong backs. That those fertile fields that now stretch to the horizon were once a wooded wilderness seems impossible to believe. To know they were cleared, one square yard at a time, picking rocks and pulling roots is difficult to conceive.

In those first years the Lays' frustrating setbacks in finding land and finally ending up, not on the Great Plains, but really in the mixed forest called northern Saskatchewan, may have been a blessing. After years of above average rainfall, the great drought and the dustbowl dissolved the dreams of many Oklahoma and Saskatchewan homesteaders on the plains. Farther to the north, the Lay homestead missed the brunt of the drought. Not only that, but other resources of the land: fish, game, and forest products helped sustain and grow the Lays' agricultural holdings. In just a decade much of the bush was cleared, gardens were cultivated, individual homes were built, and herds of cattle were fattened and sold.

One morning in the winter of 1943, Barney's youngest daughter Jeanie woke up, her nose was prickly cold, and she smiled. For the fire in the cast iron wood-burning stove to have burned so low, for the house to have such a chill, it could only mean a frigid day. For Jeanie Lay, that would mean her daddy would harness up the horses and give her and her brothers and sister, Eunice, a ride to school. Jeanie loved to ride in the wagon. At this latitude, the January walk was dark, and when the northwest wind blew from Great Slave Lake a thousand miles away, and down across Saskatchewan, it could be frigid. Just like all the other kids at school, it was all she knew. There was nothing to compare it to or complain about. It was the way it was. Even so, Jeanie was a wiry little kid and watching her lean into the wind and plant her feet with deliberate determination one foot in front of the next, brought smiles to her aunts and uncles who lived in

46

similar tidy cabins nearby. Jeanie was truly a Lay, and from early on, clearly had the tenacity of her grandfather.

Unless it was deep below zero cold, Jeanie didn't much mind the daily two-mile walk to school. She was just happy finally to be going to school. At age seven this was her first year at school. That first school, almost halfway to the town of Goodsoil, was so crowded her parents had held her back until her brother, Floyd, started school, so although almost two years apart in age, Jeanie and Floyd were in the same grade. The Lays' farm was 50 miles northwest of Meadow Lake, a half-dozen miles east of Goodsoil. Within a few years, a new one-room school was built right on the Lay Section line just a short walk from Jeanie's home. It was called Jackpine View and would be the school Jeanie would attend until grade 8.

On that frigid morning in 1943, her brothers were already out in the barn milking the cows, and she could hear her mom poking the fire box and sprucing it up. Daddy was always first to the outhouse, and Jeanie shuddered when she thought of the brisk gauntlet when the temperature was so deeply below zero.

Almost eight-years-old, Jeanie still slept in her crib. She always thought it was because she did not like sharing a bed and having her sibling's feet rub against hers. But later her mother told her that keeping her in the crib was more for her protection than for her comfort. Jeannie shared her dad's outspoken resistance to bite his lip. Small and feisty, her siblings, especially Eunice, enjoyed getting her riled up. They did so at their peril because Jeannie never backed away from a fight. Her sinewy little arm could snap like a buggy whip, and when teased or belittled she did not hesitate to throw a crab apple, dirt ball or rock at her tormentors, even if it meant a spanking. Her parents believed a little solitary confinement at night in her crib might keep the peace.

Was it a farm or a ranch? Although it would soon become a major grain producing area, in the first half of the Twentieth Century the countryside surrounding the Lays' homesteads had multiple personalities. Homesteaders were gnawing away at the forest of poplar, jackpine, and spruce. The bush was being carved into a checkerboard of farm fields. Families like the Lays spent much more time clearing the land than they did planting and cultivating crops. In the thirties, cash money was hard to come by, so the more people could grow, kill or barter, the less they needed to spend. In the dark days of winter, the men often ran traplines, and the cash from fur sales helped buy necessities from the small stores in Loon Lake and Goodsoil. Capital expenditures for mechanized equipment came slowly, and the first fields were cleared with axe, shovel, and oxen. Raising hogs and beef cattle was easier money than horticulture. Large natural hay meadows along the creeks and rivers were nourished by annual spring floods and flourished as the summer wore on. Calves fattened quickly with little more attention necessary than shooting the occasional wolf. Everyone had at least a few cattle and prided themselves on rounding them up. Even Meadow Lake town folk sometimes kept a piece of land that they might call their ranch where they kept some cattle, ran some fence and built a log cabin or two. Despite the rapidly growing fields of grain and the commercial fishing industry just to the north, Meadow Lake's identity was more linked to a cowboy culture. Dairy cows were housed in the barn, and hog pens adjoined the barn and outbuildings, but the people talked cattle. The Meadow Lake Rodeo was the event of the year, as it is to this day.

In her younger years, the town of Meadow Lake may as well have been on the moon. Jeanie's life was the homestead, even trips to the one-horse town of Loon Lake and later to the two-horse town of Goodsoil were uncommon. On the homestead, there was always plenty to eat, and similar to the walk to school, Jeanie never complained about the monotony of her diet. She found joy, as kids do, in the familiar and comforting foods which she grew up with. Porridge for breakfast seven days a week, sugar on the porridge just once in a while. Breakfast was always followed reluctantly with a teaspoon of cod liver oil. Jeanie's dad, like many of the men, boosted the family's protein by hunting wild game. Time in the bush hunting or out on the trapline meant time away from the farm. Jeanie's mom like many other frontier women was self-reliant. Despite the farming, the trapping, and the hunting, most of the men dressed as cowboys and prided themselves on their horseback and cowboy skills.

There was also a large indigenous and Metis population nearby. Mostly of Cree descent, many of these First Nation people took to the rugged and independent career of commercial fishing on lakes to the north of Meadow Lake.

The Lay family was tight and was large enough that family members could rely on other family members for expertise in everything from butchering a hog to stitching up a wound caused by the errant swing of an axe. No need for bakeries, butcher shops or grocery stores, just about everything came from the farm. Aunts and cousins often visited, usually with a basket of biscuits or muffins. The family rarely socialized much outside of the family. They didn't need to. Almost every Saturday night there was a barn dance at Vater's big barn, and talented musicians from the family played guitar, fiddles, banjos, home-made percussion instruments and Vater's new piano. Sunday afternoon was for baseball, and the Lay brothers were a formidable team. Dressed in uniforms made of flour sacks they played rough and hard and were feared adversaries in the region. Even Jeanie's dad, Barney, could play first base with aplomb, catching and throwing with his one good arm and slapping out singles between fielders almost half the time. One of the brothers, a flame-throwing pitcher, Jiggs, was scouted by a professional team.

Jeanie adored her little brother Dallas who was born in 1944. It was only six years after his birth that Jeanie began grade 8 and needed to move to the town of Meadow Lake. The town was only 50 miles away, but it was a long 50 miles, with dirt roads sometimes impassable in the spring. She boarded with relatives in return for doing childcare and helping with household chores. Motor vehicles were just coming into use on the farms, and it was as likely to see a horse and buggy on the road as a car. Meadow Lake was a small town at the time but must have felt like New York City to Jeanie. She was coming from a 3-room cabin with a dirt floor kitchen, and where major childhood outings had been bumpy trips in Uncle Patch's Model T Ford to Water Hen Lake for a swim. Despite all the diversions and excitement of town life Jeanie was often homesick. Every night, for the first several weeks in Meadow Lake, Jeanie cried herself to sleep, head buried in her pillow, missing her family. With a stiff upper lip, she put her best effort into school and paradoxically dreamed someday, not of returning to the farm, but of becoming an airline stewardess.

Jean Carter's family homestead
(photo courtesy of Goodsoil Historical Museum)

Chapter 5
Commercial Fishing

When most people think of commercial fishing, they imagine big trawlers on the open ocean. Like almost everything else, fishing in the North is different. Some of the indigenous Dene had been netting fish in northern lakes for thousands of years, but only for subsistence. The most easterly Dene groups, the Chipewyan, Yellowknives, and Dogribs, were caribou hunters and considered fish fit for consumption only in times of famine. Fishing did accelerate during the nineteenth and twentieth centuries as Anglican and Catholic missions established themselves on Great Slave Lake. Dog teams were the common mode of winter travel. Dogs were fed fish. The fish were netted in the fall and skewered on poles suspended on a fish "stage." The putrefaction that would commence on warm autumn days would keep the fish semi-pliable into the winter, and more easily digested by the sled dogs. The dogs, which seldom lived more than a half-dozen years, were incredibly tough. They could sustain themselves by gnawing on a frozen fish, and sleep curled in a ball even at fifty below zero. Dogs outnumbered the people and the evening hol was a northern cacophony. But Great Slave Lake is an inland freshwater sea of immense proportion, and the fish harvested by the Dene was inconsequential, even with the dogs to feed. Essentially, the big lake was unfished until the 1940s.

There were some early attempts at serious commercial fishing in the big lake during the summer by small boats in the mid to late 1940s. These first attempts were centered east of Yellowknife near Gros Cap. The fishing was fantastic. High-profit fish including jumbo whitefish and lake trout filled the nets. That was not the problem. The problem was getting the fish to southern markets.

During the warm summer months when the lake was ice-free, the fish had to be transported by boat across Great Slave Lake and up the Slave River, then portaged by truck across two sets of rapids at Fort Smith, and further shipped upstream by barge to the railhead at Waterways. The only practical way to do this thousand-mile journey was to freeze the fish. The fish had to remain frozen over the long wilderness transport. This taxed the technology of the day. The big market for freshwater fish was for fresh fish. The bonanza would come when fresh, unfrozen fish could quickly be transported to markets in Chicago and New York. This meant winter fishing. Winter fishing and bringing fresh fish quickly to market was not viable until there was an all-weather road from Hay River to the railhead at Grimshaw. The Mackenzie Highway, was completed in 1949 and it set off a fishing rush similar to a gold rush.

It is possible to manage a commercial fishing industry so that it is sustainable. In theory, fishermen harvest the surplus fish annually and leave enough fish for reproduction and a stable population. To get it right there are many variables to consider including the size of the fish harvested, when and how they are caught and the interplay with other species. Initiating regulations involves trying to understand the science of it all, and not letting politics and short-term economic motivations muddy the water. Effective resource management also assumes the scientific community has it right. When in truth, especially in 1949, there was nothing more than a broad consensus that conservation should be encouraged. Data driven recommendations were not reliable. Regulations were based on what little was known. Much was not known. The independent spirit of northerners and their innate distrust and frustration with regulations imposed from "outside" made adherence dodgy.

All this is just tavern talk. What was real in 1949 was a vast freshwater lake, the ninth largest in the world, teeming with highly esteemed fish. It was a lake virtually untouched and finally linked to the North American market by an all-weather road. The other truth was that there were fishermen skilled in their trade making a living fishing more southerly freshwater lakes, but facing diminishing returns as these more southerly lakes became depleted of fish stocks. News and rumors of the incredible opportunity of Great Slave Lake created a fever that spread like a wildfire through the freshwater fishing camps across the Great Lakes, Lake Winnipeg, and northern Manitoba and Saskatchewan. These fishermen knew the first few years of fishing would be the best. The yields could be fabulous.

Fortunes could be made. That urgency, excitement, and those green eyes further stoked the fire. The urge to get up there now, and set the first nets, was fever pitched.

Two years before the road was even completed four companies began establishing commercial fishing operations in Hay River. In 1949 when the Mackenzie highway was opened, Carter Fisheries was one of the original 13 companies to be based in Hay River. Although George Carter likely had the least financial backing, Carter Fisheries thrived, while six other fish buyers failed. Ten years later, Carter Fisheries was one of the big three of the seven surviving companies. In appearance, with smartly painted buildings and neat sheds, even if it wasn't true, it seemed as if Carter's was the biggest. George had earned the reputation as an austere but fair employer, and his workers were loyal. He had the panache to be equally at ease drinking a cup of tea with a Metis fisherman as sipping a cocktail with a Jewish fish buyer in New York City.

The fishing industry was run like an old mining town. Companies brought in crews and housed them, fed them, equipped them, and took care of them. In return, the fishermen caught the fish, bought items at the company store and were loyal to their company. Fishing was seasonal, and credit was extended to fishermen during lean times by the company store and repaid during the fishing season. A cynical viewpoint would be that the fishermen were loyal to "their" company because they owed their company money. Most of the fish companies like Carters preferred independent fishermen. George encouraged loyal fishermen that had put in a couple of seasons of reliable work to become independent. Carter's would front them a year's salary ($3,000 in 1960), and they could buy their nets, Bombardier, and supplies. Rather than receive a salary, they could sell their fish to Carter at market prices. When airplanes were used to transport the fish, pilots would separate each crew's catch in the airplane's cabin with tarps of canvas. When unloaded and weighed the pilots were given a chit to take back to the fishermen on the next haul. Once they paid off their loan, they could sell their fish to whoever they wanted. In Hay River most fishermen (and pilots) stayed with the same company. There was little acrimony between companies. With few exceptions, companies helped each other out, and if a bug (Bombardier) broke down crossing the lake, the owners of the bug needed only walk to the nearest light to be assured of assistance, regardless of the company.

George's familiarity, belief, and early embrace of aircraft as an integral part of Northern winter fishing was one factor that helped push Carter Fisheries to near the top of the pile. The best fishing is in the early part of the season, just after shore ice is thick enough to hold men and equipment (four inches of good ice). The Mackenzie highway was built to Hay River, but there were no networking roads suitable for trucks along the shorelines. The best fishing was also across the bay from the town. Great Slave Lake was not quick to freeze over. Ice, even thick ice, could extend miles out from the coast, but there would still be open water between the fishing banks and Hay River. The big lake would "try" to freeze over, but the north wind would kick up and smash the ice to shattered icicles. Air transport was the only way to get the early season catch to Hay River and the road south. Other companies used aircraft too, but Carter's use was the most aggressive.

Although George was a stern and phlegmatic boss, he was not without compassion. He was not beyond extending credit to loyal employees who were having hard times. He also had a reputation of delivering a fresh and excellent product to fish buyers in New York City. Just as the Lay clan was carving a farm out of the wilderness, pushing plows and pulling stumps with muscle and powered by sweat, George was building a fish empire with guile, risk, and stern business savvy. The traits both families shared were faith in what the future would bring and an unwavering determination for success.

Fish that might have been acceptable to other companies would not pass muster with George. And George seemed to be everywhere. With hands in his pockets, George would check fish as they were unloaded from the planes and he would insist anything less than top restaurant grade be discarded. He seemed to be looking over the shoulder of every truck driver to make sure the fish were packed correctly. George also had the ability to see talent in an individual that someone else might overlook. Early on, he recognized not only his son's love for aviation but his talent for it.

George's pilots would use his Cessna 170B and later his Cessna 180s to get the crews out as early as Remembrance Day, November 11. Legally, with a private pilot's license, Merlyn could carry commercial freight, but not people. So until 1956, his passengers were primarily fish. On that thin early season ice, planes barely stopped long enough to pick up a quick load and then fly off again. The goal was to start hauling fish as early as possible without breaking through the ice and going into the soup. Some years that was later in November and some years, by regulation, December 1.

Going through the ice was a real possibility. Decades later Merlyn will break through with one of his Single Otters on a fish run. The wings will keep the plane from sinking, but it will be a big job to extricate the fuselage, repair the damage, and install a new prop, before he can fly it out. Another of Merlyn's pilots his nephew, Stu Poirier, will also put an Otter through the ice on the east arm of Great Slave Lake.

Jones' Point, Great Slave Lake, IOF through the ice
(photo courtesy of Carter family album)

Every season the 170B and the 180s were the first planes used to haul fish. Loaded, the 170B weighed a bit over a ton, and the 180 a ton and a half. Pilots preferred 6 or 7 inches of good ice to safely land and load but sometimes had to settle for as little as 4 inches. Most of Carters' fishermen came from the Meadow Lake area and the lake region just to the north of Meadow Lake. George also hired crews from Gimli, Manitoba. Gimli is a small town on the western shore of Lake Winnipeg. The area was originally settled by Icelandic fishermen who, in 1949, were still fishing Lake Winnipeg, but with diminishing returns. Many of these fishermen were also skilled carpenters, especially skilled working with thin marine plywood. Everything about hauling fish in an airplane is about load. Every ounce counts. The place for airplane seats was back in the shed right next to the ten-pound radio. Fish boxes were for trucks. Planes were stripped clean. In the off-season, the Icelandic carpenters would line the interior of the airplanes with thin marine plywood and then shellac the wood. At the belly, they would install an integral funnel with a drain hole. The shellacked plywood interiors could be swished out with a bucket of hot water at the end of the day. Even so, on one occasion the flap lever for one of Carter's Cessna 180s jammed. When the floor boards were pulled up, the control cables were discovered to be encrusted with frozen fish slime.

Depending on the plane, sometimes fish were loaded through the window. A small partition of plywood surrounded the pilot, holding back the loose fish and preventing them from sliding into the rudder pedals. Other than that, fish just filled the plane without concern whether fish sloshing around might affect the aircraft's weight and balance. When the door was opened to unload the plane, the fish spilled out like silver dollars from an old slot machine. On straight skis, the Cessna 170B with just a little 145 horsepower engine could haul 600 pounds of fish. A 180 could carry about 1000 pounds. These planes, and similar planes from other fish companies were running back and forth from the fish camps on the big lake to the West Channel non-stop from dawn to dusk.

These 30-mile "milk runs" were not without potential for mishap. On rough fields, a tailwheel airplane such as a Cessna 170B has a superior gear configuration to a modern tricycle (nose-wheel) aircraft. Like a wheelbarrow that hits a woodchuck hole, landing a nose-wheel plane in meadows or gravel bars creates a propensity to flip the airplane, or snap off the nose-wheel gear and plow into the ground. Tailwheels would harmlessly bounce or roll over a similar obstruction. In the winter, skis can be easily fitted to replace the main wheels on a tailwheel configured airplane. On a 170B, the tailwheel is replaced with the blade of a shovel. Ski-equipped in the cold dense air, the 170B was an excellent performer and the 180 was exceptional. Because of its all-metal construction, big barn door flaps, long coupled gear, broad wing, and Cessna replacement parts, the 170B is a highly sought after "vintage" plane to this day. Often described as underpowered in current literature with recommended bigger engine modifications, Merlyn always maintained that the original 170B equipped with a 145-horsepower engine had plenty of power if you knew how to fly it and let the wing do the work and not push the tail up before it was ready, or try to jerk it off the ground relying on horsepower. Merlyn felt the same way about the Cessna180 and did not like the 185. He claimed the 180 could carry the same load as the 185 with lower fuel cost, less maintenance, and a quieter ride.

Even though it was one of his all-time favorites, the 170B snake-bit him. In 1953, in fading light and after dumping his load of fish, Merlyn, still a teenager, was taxiing his plane when he hit a wind-packed snowdrift. The front skis stuck and the inertia lifted the tail up and over and put the plane on its back. The mishap was not because he was moving too fast. When steering a nose-wheel airplane on tarmac, it is easy to make slow tight turns. The rudder pedals control the nose-wheel. Also, by selectively applying the brakes the plane can be spun and stopped tight and slow. This is not true with a taildragger on skis. In many different snow conditions, the pilot must keep the plane moving briskly just to make sure it does not stick. Turning is dependent on prop wash over the tail rudder. This takes a lot of power. Big sweeping turns at speed is the correct protocol. The 170B was quickly repaired, but one can only imagine Merlyn to be sheepish to crash while taxiing on the snow. It would be almost two decades before Merlyn would flip another plane on its back.

Depending on the weather and how fast the cold was making ice, the "milk runs" with the small planes would last only a few weeks. By then, there would be a foot or more of solid ice at the camps, and it would be safe to bring in the heavier planes. Although the camps near shore had good ice, there would still be open water and thin ice between the camps and Hay River. Planes still ruled. The Bombardiers with their attached caboose trailers had to sit idle and snow-covered at the West Channel waiting for the fierce arctic winter to make deep ice.

By the end of December, the frigid air had won the battle against water. Western bays of Great Slave Lake were frozen solid. As soon as the ice was safe, Bombardiers took over the fish hauling. Bombardiers were snow machines often called "bugs" that looked like art deco flying saucers. They had two tracks in the back for traction and two skis up front for steering. They could carry and drag a caboose (a sled capable of a heavy load) of fish far more economically than aircraft.

Bush flying is almost always a temporary solution until a more cost-effective mode of transport is available. Whether it be fish or hauling freight, if there is time and open water, barge transport is the cheapest. If there are rails, trains are the second most economical way to move freight. Then comes trucks if there are roads, even winter roads. Then comes cat (bulldozer) trains across lakes when the ice is deep or pulling the train along cut lines through the bush if the muskeg is frozen solid. Next come the Bombardiers with their cabooses, heavy but much lighter than cat trains. Finally, airplanes runway to runway, then more rugged airplanes dirt strip to dirt strip. Bush planes on floats or skis are the most expensive on the list. Bush planes are the pioneers, the first to arrive and the first to haul freight. Like all pioneers, their time is usually short lived. At the forefront of exploration and development, their usefulness hinges on a dearth of more economical means of transport. That hasn't changed much since 1949, although helicopters could be added to the bottom of the list as the most expensive cargo haulers.

Driving heavily loaded Bombardiers across Great Slave Lake was not without its perils. Once the lake was solidly frozen, warm spells would cause the ice sheet to swell and push up pressure ridges. Smaller ridges were obscured when the wind blew across the lake and Bombardiers hitting a small ridge at high speeds sometimes threw a track or broke a front ski, or jackknifed and rolled. Those collisions would spill hundreds of fish and a few fishermen on to the ice. Bigger ridges were easier to see, but sometimes impossible for the bugs to climb over. Fishermen would need to get out of their bugs and hack out a notch in the ridge with axes, needle bars, and shovels. The notch needed to be big enough and low enough for both the bug and caboose to pass through.

More harrowing for fish transport were during cold snaps. When the temperature dropped to 40 below zero, and colder, the lake's ice sheets shrunk back, pulling apart at weak spots and creating cracks of open water as much as 12 feet wide. Fishermen needed to work fast and transport the fish across the lake before the fish froze. Fishermen would sometimes shovel snow into the cracks to create slushy fords, but that took time. More likely they would just get a running start, rev their bugs to maximum rpm, and hydroplane their loads across the crack the way a child can skip a stone across a pond. At 40 below zero, there was no margin for error. The fishermen of Great Slave were highly skilled at their craft. Eugene McKay, a Cree, was a best friend of Merlyn. He had learned the trade of fishing on Saskatchewan lakes north of Meadow Lake. Although he could not read or write (he carried a rubber stamp of his name in his pocket), Eugene was an astute businessman and managed a successful crew of fishermen on the big lake for many years. He never lost a bug or a load of fish, but he was careful and knew from the weather conditions and the lake morphology where and how cracks and ridges would develop and how best to navigate them.

Despite the cracks and the ridges, when the lake ice became thick enough all the way across the western end of Great Slave Lake (a solid foot of clear ice), Bombardiers could start hauling the fish and it ceased to make sense to fly fish through the air when they could be hauled across the ice. The role of aircraft changed with the seasons.

Almost as soon as ice was thick enough for the Bombardiers to start hauling the fish directly from the camps on the western end of the Great Slave Lake to Hay River, fishermen were catching fish in the Simpson Islands on the Great Slave Lake's eastern arm.

For the first couple years George, and initially in partnership with his brother Lawrence, hired Stan McMillan and his Fairchild 82 to do some of his fish hauling. In 1950, even before Merlyn had obtained his private pilot's license, George put Merlyn in the right seat as co-pilot. Merlyn credited that time with McMillan as his most precious bush pilot tutelage.

Avro Anson at the West Channel (courtesy of Carter family

George hated hiring someone else to do his flying, a trait he passed on to Merlyn. As soon as he could arrange it George bought his first big plane at a military auction. The Avro Anson Mk V was a Canadian-made warbird. It was a low-wing taildragger with twin radial engines. It was made of wood, and although built for military navigation and training purposes, it did have bomb doors. Merlyn eventually wired them shut, presumably, so he wouldn't lose a load of fish or a passenger. It was not a great fish hauler because the small passenger door made loading and unloading fish awkward. The low gear and resulting minimal prop clearance and low wing, made landings on rough chunky ice or rough strips challenging. If the wooden props hit a chunk of ice, they would splinter. Even so, first with Merlyn in the right seat and then the left, for almost a decade, Merlyn hauled many tons of fish in the Anson.

During this phase of the winter fishing season, the smaller planes did not sit idle. They were used to check ice depths at new camps before bringing in heavier aircraft. The small planes were also used to move crews around. George could also utilize the planes to start establishing fishing camps on the eastern arm of Great Slave Lake, over 100 miles away.

Great Slave Lake is a massive lake and the ninth deepest in the world. Over 100 miles to the east of Hay River at the beginning of the Canadian Shield, huge stocks of lake trout and jumbo whitefish plied its waters. George had started establishing winter fish camps on the Simpson Islands as soon as the narrows started freezing up solid. Despite weeks of frigid weather, a lake the size of Great Slave takes a long time to freeze to a depth that makes it safe for Bombardiers to carry and pull tons of fish. Until lake ice was thick and stable, aircraft were needed to haul the big catches to Hay River. As soon as possible George sent the big planes out for the long haul into West Channel.

By Christmas or before, the entire Great Slave Lake was frozen hard. Bombardiers and in some cases trucks could then haul fish all the way from the Simpson Islands to Hay River. Again, aircraft were superseded by slow cheaper ground slugs, the bugs.

After a couple of years, what had first seemed like an endless supply of fish from Great Slave Lake had already started to taper off. There was also a time period in mid-winter when ground transport could handle all the loads. George did not like to see his planes sit idle. He looked to the big interior lakes of the Territories that up to that point had been unharvested. Lakes far from roads or cut lines, lakes that could only be serviced by aircraft.

Not every interior lake had great fishing. Not every interior lake had the species of fish in demand by fish buyers. In the business of aviation, fuel and maintenance expenses gnaw away at the profit margin. The difference between flying fish worth ten cents a pound and flying fish worth a nickel a pound was the difference between living high and going broke. Setting up crews and camps on lakes where the right fish lived was critical for success. It was also critical to minimize the distance the fish needed to be transported to Hay River, especially if that transport was to be done by aircraft.

Long before winter fishing season began, back in the summer when the sun just rolled around the sky like a marble in a cereal bowl, George started to send his son out fish prospecting in the float-equipped Cessna 170B. With a short gill net, sinkers and floats, an old Enfield rifle, some grub, maybe a little canoe tied to the pontoons, Merlyn flew east and camped on O'Connor, Rutledge, Gagnon, Thuban, Thekehili, Sparks, Porter, Tsu, Artillery, Lady Gray and King, lakes occasionally even as far as Nonacho and Gray Lake. He was looking for jumbo whitefish, and he found them. Carter Fisheries would later put winter crews on many of these lakes, but some, Nonacho in particular, had too many huge lake trout. Fish upwards of 50 pounds that would spin in the net and twist it into a rope. Some like Gray Lake were too far and the time and the fuel cost would chop away at the profit margin like a voracious shark. Merlyn learned what species was in which lake, and he learned where on the lake was the best place to set a net. He had the best success locating nets at a narrows, or an intake or outflow. He specifically remembered Nonacho where the trout were so big he almost upset his canoe checking the net. Merlyn also prospected for fish to the west, checking Tathlina, Kakisa, Trout and Dogface Lakes. He found pickerel (walleye) in abundance but not many whitefish. He also checked a myriad of lakes north of Yellowknife including Point, Pellat, Contwoyto, MacKay, even Great Bear itself.

But there was one lake that outdid the rest. It was Hottah Lake just southeast of Great Bear Lake. It was possible on Hottah Lake to pull 400 jumbo whitefish out of a single net. This was an astonishing total.

Hottah Lake, however, garnered fame for two other reasons. It was at the epicenter of a wave of mining rushes in the mid-twentieth century. Uranium for the bombs dropped on Japan was mined nearby, and several small gold mines were also developed in the area. There were two mine sites on Hottah Lake itself.

Merlyn up top fueling the Cessna 170B
(courtesy of the Carter family album)

George recognized not only the astonishing fishing in Hottah Lake for the prized jumbo whitefish but also the possibility of using some of the abandoned mine buildings as fish camps. Hottah Lake is one of the most frigid locations in North America. Surface ice on Hottah Lake could reach seven feet deep, and the lake typically remained frozen well into July. Two-hundred-and-eighty miles from Hay River it was too far to use aircraft economically as primary transportation. Every winter an ice road suitable for truck travel was built from Yellowknife to the mines near Port Radium on Great Bear

Lake. George correctly surmised that fish could be flown from Hottah Lake to lakes along the ice road and then transported the rest of the way to Hay River by truck.

Hottah Lake is also ignominiously remembered because of an epic saga of survival. This anecdote may illustrate just how remote the countryside was during Merlyn's entire lifetime.

In November of 1972, a bush pilot named Marten Hartwell was flying a medivac from Cambridge Bay with three passengers in a Beech 18. He encountered bad weather and lost his orientation in cloud before crashing into a hillside adjacent to Hottah Lake. A nurse was killed on impact; a pregnant Inuit woman died soon after, but Hartwell and an Inuit boy, David Kootook, who was suffering from appendicitis, survived. Hartwell's ankles were both broken by the crash, along with his left knee and nose. For weeks, the two survived arctic winter weather and an average temperature of −35 °F. Kootook's Inuit campcraft skill was vital to the survival of the pilot. Kootook built a shelter, kept a fire going, ate edible plants and cared for the pilot. Finding anything edible on the barren lands in winter is a challenge that defies comparison. Sadly, Kootook died of starvation after the 20th day. The pilot survived by eating the nurse. Inuit, like many cultures that live in areas where famine is always a possibility, have a strong taboo against eating human flesh. Hartwell did not share that taboo, but it became evident he was uneasy of the deed. Hartwell greeted his rescuers after 31 days in the frigid wreck, by saying, "Welcome to the camp of a cannibal." He also refused to sell his story to the media. Kootook was honored posthumously for his bravery and resourcefulness. It is impossible to empathize with either individual's choices unless you have spent three weeks marooned on the frozen tundra.

Although readers might be flabbergasted that it took 31 days of searching to find a downed silver airplane on wide open terrain, it might illustrate just how vast the Canadian North is. In 1972 the Northwest Territories dwarfed even Alaska in size.

In area, the Northwest Territories was 15 times the size of Minnesota or over a half million times bigger than Manhattan. To put it in a population density context a tourist would be more likely to bump into 2.4 million different people in Manhattan before a tourist bumped into one in the Northwest Territories. When the capital city, Yellowknife, is taken out of the mix, the backcountry of the Northwest Territories was, and is, virtually uninhabited.

Perhaps an even sadder story than the Hartwell saga was the demise of one of Merlyn's best friends. Chuck McAvoy crashed in his Fairchild 82, northeast of Hottah Lake in June of 1964. Despite an extensive search, that included Merlyn and many of the pilots mentioned in this book, it was 34 years before the wrecked airplane and bodies of McAvoy and his passengers were found.

It is impossible to exaggerate how remote, empty and vast the Northwest Territories was and in many ways still is. In his lifetime, Merlyn flew more than three million miles and spent over 25,000 hours flying over this uninhabited and wild landscape. Often, his only company was dead fish.

Cessna 170B
Data from Jane's All The World's Aircraft 1955–56
General characteristics
•Over 5,000 Cessna 170s were built
•Crew: 1
•Capacity: 3 passengers
•Length: 24 ft 11 ½ in (7.607 m)
•Wingspan: 36 ft (11 m)
•Height: 6 ft 7 in (2.01 m)
•Wing area: 174 sq ft (16.2 m^2)
•Empty weight: 1,205 lb (547 kg)
•Gross weight: 2,200 lb (998 kg)
•Fuel capacity: 42 US gal (160 L; 35 imp gal)
•Powerplant: Continental C145-2 air-cooled flat-six, 145 hp
Performance
•Maximum speed: 140 mph (225 km/h; 122 kn)
•Cruise speed: 120 mph (193 km/h; 104 kn)
•Stall speed: 52 mph (84 km/h; 45 kn)
•Endurance: over 4.5 hours
•Service ceiling: 15,500 ft (4,700 m)
•Rate of climb: 690 ft/min (3.5 m/s)

Cessna 180
Data from Cessna
General characteristics
•Crew: one
•Capacity: five passengers
•Length: 25 ft 9 in (7.85 m)
•Wingspan: 35 ft 10 in (10.92 m)
•Height: 7 ft 9 in (2.36 m)
•Wing area: 174 sq ft (16.2 m^2)
•Empty weight: 1,700 lb (771 kg)
•Gross weight: 2,800 lb (1,270 kg)
•Powerplant: 1 × Continental O-470-U , 230 hp (170 kW)
•Propellors: 2-bladed constant speed, 6 ft 10 in (2.08 m) diameter
Performance
•Maximum speed: 148 kn (170 mph; 274 km/h)
•Cruise speed: 142 kn (163 mph; 263 km/h)
•Stall speed: 48 kn (55 mph; 89 km/h)
•Range: 890 nmi (1,024 mi; 1,648 km)
•Service ceiling: 17,700 ft (5,400 m)
•Rate of climb: 1,100 ft/min (5.6 m/s)

Avro Canada Anson Mk II
Data from http://www.bombercommandmuseum.ca/anson.html
General characteristics
Engines: 330 hp Jacobs L-6MB
Wingspan: 56 feet 6 inches
Length: 42 feet 3 inches
Maximum speed:188 miles per hour
Cruising speed:155 miles per hour
Service ceiling:16,200 feet
Rate of climb:870 feet per minute
Maximum weight:7650 pounds
Empty weight:5880 pounds
Performance
•Empty weight: 6,693 lb
•All-up weight: 9,460 lb
•Maximum speed: 190 mph
•Cruising speed: 145 mph
•Initial climb rate: 1,500 ft/min
•Service ceiling: 21,450 ft
•Range: 580 miles

Chapter 6
Sport Fishing?
A Love Story

Jean
(courtesy of the Lay family album)

Pee Wee Reese is at the plate. Bob Turley throws a low inside fastball, but the quick stepping Reese is able to lay down a bunt. Turley makes a nice play, and despite a roar from the Brooklyn crowd, Reese is out in a close play at first, and Gilliam advances to second. Bottom of the 10th inning, in a scoreless tie. Two down, National League Home Run Champion of the year, Duke Snider is up to bat. Yogi Berra steps to the left to catch four consecutive pitches, and Snider trots to the first base bag, an intentional walk. Still two down, Gilliam on second, Snider on first, Jackie Robinson at the plate. Line drive to left, Mickey Mantle can't get it, Slaughter can't get it. Gilliam rounds third; he's going to score. Dodgers win. Ebbets Field is going wild. The 1956 subway World Series is all tied up. Yankees, Dodgers: three games apiece.

Merlyn pulls Jean Carter to his chest, "We're not in Meadow Lake anymore!"

A two-week honeymoon in New York City includes Broadway plays, Judy Garland at the Radio City Music Hall and the World Series. When those Carters did something, they never did it halfway. They also scheduled their wedding early in October so that Merlyn could be back North in time for the start of the winter fishing season. That was their way; business trumped family dinners, parties, and vacations. But when they were together, they were together, and they spared no expense of money or time to make the most of each other.

Merlyn had always been a slender and fit young man, but in the summer of 1954, he ate a lot of popcorn. He lingered around the Lux theater in Meadow Lake every night that August. Jeanie noticed but wasn't so sure about him. He never seemed to have his feet on the ground. After all, she wasn't starved for attention; there was Tommy Spence, Buddy Sweeny, and Herbie Price. When Merlyn finally asked her out for a date, a drive to a nearby tower where the young people hung out, she wasn't so sure. After talking the invitation over with her co-workers at the Lux and hearing them encourage her to go, she finally accepted. The first date at the tower was at best unremarkable. Jeanie refused to join in when Merlyn drank beer. He left her on the ground when he climbed the tower and acted "goofy."

And when Merlyn tried to get a little fresh, Jeanie grabbed both his hands and said, "Mister Merlyn Carter, I am not that kind of girl." Nevertheless, he kept buying popcorn. She kept accepting rides. Merlyn was to share with me forty years later that there was never another woman, before or after Jeanie, that touched his heart and filled his soul as she did. Merlyn seldom talked that way, so it was something I will never forget. He went on to say, "I knew it the moment I met her."

Jean was always in perpetual motion, bouncing like a droplet of water dancing on a hot skillet, house cleaning with an efficient fury, cooking and baking with frenetic energy, joking and storytelling with the edge of a barber's razor. The only time Jean slowed down and completely relaxed was when she leaned back into Merlyn's arms, her face smooth, her eyes dreamy, her slender arms soft, her voice tender.

The only-child who hid his moody side and the feisty sibling who fought for everything she had. The two were flint and steel, and when they came together, there was sometimes the warmth of fire and sometimes just wild sparks.

Their first home was an 8-foot by 20-foot trailer parked at West Channel. The water came from the river in a bucket. In the summer, the settlement at West Channel was either muddy or dusty. In the winter it was cold, often with a raw wind roaring off the frozen lake. Merlyn spent much of those first winters gone, from before dawn to dusk, seven days a week. For Jean, the accommodation was no smaller or less appointed than what she grew up with, and she was in love. When her man came home smelling of fish and with fingers blistered from the cold Jean was happy. The two would play a romantic game of hide and seek. In that tiny trailer, it could not have taken Merlyn long to find Jean. Especially, as she wanted to be "found" as much as Merlyn wanted to "find" her. Hide and seek was a game they would continue to the end. It was also the only time they would let the phone ring unanswered.

But Jean missed her family dearly and had to cope with in-laws that were both in-laws and employers. There is also nothing quite like a mother-in-law of an only-child. There were sparks too with

Merlyn. When he came home late from drinking and carousing with his pals, Jean was not so forgiving. Merlyn, when he opened the door to his home, would more likely be greeted by a flying cup or saucer, than a kiss. After a night of drinking Merlyn once asked a buddy to open the door, and sure enough, a flying plate hit his friend square in the head. Jean would throw dishes, sputter and might shake her head, but not for long, and soon would be joking and laughing with Merlyn. Usually summing up, that men are just confounding partners with the cliche, "Can't live with them, can't live without them." Merlyn was also fond of sharing that in his house he always got in the last two words, "yes, dear."

As for Merlyn, he was good at dodging dishes, omitting details of antics and flying under Jean's radar. "Oh jeez Don't let Jeanie see that one, boy, I could get into big trouble over that one, you betcha." and then with a big smile a laugh that would cause his stomach to bounce. He seemed always to know that no matter what he did, eventually she would forgive him. There was no time to stay mad. The fish business was booming, and the flying was busy. At the time of their marriage in 1956, the Carters already owned three airplanes, the Anson, the Cessna 170B and the new Cessna 180 KOW. Merlyn was doing most of the flying.

Bound by the restrictions of a private license, it was imperative that Merlyn obtained a commercial license as soon as possible. Although his flying skills were already outstanding, the bookwork was a stumbling block. Merlyn's reading abilities were limited, and Jean read and re-read manuals and textbooks out loud until Merlyn was finally able to learn the theory and the regulations necessarily to pass the written commercial pilot exam in Winnipeg. Twenty-five years later, when his son Myles kept struggling unsuccessfully to pass the written portion of his commercial pilot's exam, Merlyn often confided that he didn't think he could even pass the current test. Merlyn was never much for reading. I remember flying with him in 1976 and stopping at Norman Wells. The weather was a little iffy. We came into the flight service area to use the bathroom and on our way out Merlyn asked the controller what the weather was doing. The guy just pointed to the weather Telex over in the corner and said, "Take a look." I thought we would, but Merlyn aimed for the door and off we went.

It would be a mistake to underestimate Merlyn's business acumen by judging his reading skills. Merlyn freely confessed that he did lean on Jean to do the essential paperwork of the business. Just two months before his death, when inaugurated into the Northern Air Transportation Association Hall of Fame, he gave Jean credit for doing all the hard work, the business end of aviation. He claimed he did the fun stuff — the flying. Jean would be the first to disagree, and she would be right. The success of Carter Air Service depended on the two of them working together. Merlyn did the flying and made the big business decisions. Jean kept the books, the schedules, and the payrolls. Together, right from the start, they charmed the customers and answered the phone on the second ring, "charter service anywhere."

Just two days short of their first anniversary, Jean gave birth to their first son, Dean. Jean had gone 700 miles south to Edmonton a month earlier to wait for the birth. Jean's blood type is Rh-negative, and Hay River did not have the health care capabilities to handle a potentially difficult birth. She had a comfortable place to stay in "the city," as Edmonton was called by northerners. Her parents had moved from the family homestead to Edmonton a few years earlier because Barney was having serious problems with his heart. An October due date was good for Merlyn as October is a slack season for flying in the North, too cold for floats, too warm for skis. So Merlyn was with Jean on the morning of October 4, 1957. It was an early season snowstorm when the time came for the run to the hospital. Traffic was snarled, and they were stuck on the middle of the High Level bridge in Edmonton. Jean was in the back seat in full labor and sick to her stomach and Merlyn was sitting behind the wheel with the car at a standstill; both the newlyweds were crying. Merlyn found a nail file and was sharpening a shoehorn, "What cord am I supposed to cut?"

They made it to the hospital with seventy minutes to spare. The shoe horn could go back into Merlyn's pocket. Two years later Kandee was born in January, fish flying season, and Jean was on her own for that one. Part of the life of a bush pilot is not making it home every night. There were contract hauls, game surveys, and mineral staking ventures that could keep Merlyn away from home for weeks at a time. Bush pilots had to be ready to fly at a few minutes' notice, and willing to be away from home, sometimes for extended periods of time, maybe sleeping in the back of the airplane, a tent, or a seedy northern hotel. When you are in business for yourself, and you are a pilot, no matter what is happening at home, you have to be in the air, or you will go broke. The fact is, the only thing that kept Carter Air Service from going bankrupt was the dramatic inflation in the value of bush aircraft. Every time Merlyn upgraded to a new plane, he sold his old plane for more than he had paid for it. Sometimes as much as three times more. Merlyn often remarked that between the weather, the high cost of fuel, maintenance, insurance, damaged planes, more and more regulations, that as much as he kept his planes in the air, it was hard to keep his head above water. Merlyn would say running a northern aviation business was like filling a bucket with water, but the bucket had lots of holes. The water runs out almost as fast as you fill it. The worst part was just when you have the bucket almost full; something kicks it over and you start all over again.

The biggest aggravation for Merlyn was meeting all the government regulations. Many times, the regulations almost drove the Carters to close it all down. As much as he loved the flying, often he thought seriously of selling the business, including once when Merlyn came so close to closing a deal with Perry Linton that Merlyn and Jean had already sold their West Channel home. This was the same Linton who a decade earlier had made the first flight of his life with Merlyn in the 170B.

That sale fell through, and Merlyn's only daughter, Kandee, grew up in their big Vale Island house and was never quite sure when or for how long her dad would be home. Kandee still loves the smell of raw whitefish. She remembers waking up to that smell when she was a little girl and knowing that her father had come home. That always made her happy. Although not that involved in her care, Kandee was always the apple of Merlyn's eye, and Kandee had an unwavering devotion to her father.

The early sixties were pivotal years for the Carters. After one of George's hired pilots, Bobby Maclean, crashed the Anson on take-off in 1961, Merlyn replaced it with a Beech 18 and a Lockheed 10A. Both planes were dedicated fish haulers, but the Lockheed with its big soft tires and higher gear could carry a much heavier load on the makeshift lake-ice landing zones. The Lockheed 10A was thirty years ahead of its time, and it certainly became a favorite with Merlyn. Fish harvests were still strong. Daylight is precious during a northern winter and both airplanes and two crews were kept busy all winter from first light to complete darkness. Merlyn and Dallas Lay (Jean's brother) flew the Lockheed and Jim McAvoy and Joe McBryan were in the Beech. Those were also winters when fish prices were stable and overhead was increasing. Always perceptive, Merlyn sensed that soon a time would come when fish flying alone would not support his growing family. If Merlyn was to keep the business, he needed to find ways to get his airplanes up in the air, especially in the 24-hour sunlight of summer. Merlyn had always utilized his planes to do some fuel hauls and move trappers and prospectors around, especially so after earning his commercial license in 1957, but summer flying had not been a major focus.

With Dean and Kandee as toddlers and Myles soon to come, Jean had temporarily stepped back from involvement in the business and dedicated herself to domestic duties. Jean confirmed that Merlyn lived his entire life without ever touching a diaper, a dishrag or a mop. He also left the childrearing to his wife. In his teens and early twenties, the summer had been a time for vacationing in Meadow Lake, prospecting for fish in the interior lakes with the float equipped Cessnas and for travel around Canada and the States. But gears in his head were turning. As early as 1960 he was working toward a forming a year-round charter service, and in 1962, Carter Air Service was officially incorporated.

There was a geographical problem with an air charter company based in Hay River. Hay River was nicknamed the "Hub of the North" because of three transportation developments. The Mackenzie Highway connected the town to the south in 1949. In 1964 a rail line from the South was built and connected the town by train. Most of all, Hay River was the base of the Northern Transportation Company Limited (NTCL) and was the major staging point for tugs pushing barges down the Mackenzie to its delta and beyond, including arctic Canadian settlements and Alaska's north slope. From a charter aviation standpoint Hay River was a hub in a negative way. At the end of a spoke to the west was Fort Simpson, already a shipping destination from both the Liard and Mackenzie Rivers, and the town would be connected to the highway system in 1970. Yellowknife, at the end of a spoke to the north, was on the highway system soon after Hay River and had a large airport. Fort Smith, at the end of a spoke to the east, was the historic Northwest Territories capital and had a developed infrastructure and modern airport. To the south, there was High Level, Alberta. All of these communities were better and more logical and economical starting points for charters down the Mackenzie valley, to the Arctic, or to the caribou range. A Hay River based charter company would always be at a competitive disadvantage for lucrative long-distance contracts. Merlyn sensed that if he were going to build a successful charter business based in Hay River, he would need to create his market. By the 1960s, veterans of the Second World War were in established careers and their kids were growing up. The men and in some cases their wives were looking for adventure. Nothing spells adventure like virgin lakes and giant trout. Merlyn knew where they both were.

*Merlyn's Lockheed 10A Electra
(courtesy of Carter family album)*

Lockheed 10A Electra
Data from Wikipedia
General characteristics
•**Crew:** two
•**Capacity:** ten passengers
•**Length:** 38 ft 7 in (11.8 m)
•**Wingspan:** 55 ft 0 in (16.8 m)
•**Height:** 10 ft 1 in (3.1 m)
•**Wing area:** 458 ft² (42.6 m²)
•**Empty weight:** 6,454 lb (2,930 kg)
•**Loaded weight:** 10,500 lb (4,760 kg)
•**Powerplant:** 2 × Pratt & Whitney R-985 Wasp Junior SB, 450 hp (340 kW) each
Performance
•**Maximum speed:** 202 mph (325 km/h) 175 knots
•**Cruise speed:** 190 mph (306 km/h) 165 knots
•**Range:** 713 mi (1,150 km) 620 nm
•**Service ceiling:** 19,400 ft (5,910 m)
•**Rate of climb:** 1,000 ft/min (300 m/min)

Beechcraft 18 (Twin Beech)
Data from Jane's Fighting Aircraft of World War II.
several versions the Beech 18 were built
General characteristics
•Crew: 2 pilots
•Capacity: 6 passengers
•Length: 34 ft 2 in (10.41 m)
•Wingspan: 47 ft 8 in (14.53 m)
•Height: 9 ft 8 in (2.95 m)
•Wing area: 349 ft^2 (32.4 m^2)
•Empty weight: 6,175 lb (2,800 kg)
•Loaded weight: 7,500 lb (3,400 kg)
•Max. takeoff weight: 8,727 lb (3,959 kg)
•Powerplant: 2 × Pratt & Whitney R-985-AN-1 "Wasp Junior" radial engines, 450 hp (336 kW) each
Performance
•Maximum speed: 225 mph (195 knots, 360 km/h)
•Range: 1,200 mi (1,000 NM, 1,900 km) at 160 mph (260 km/h)
•Service ceiling: 26,000 ft (7,930 m)
•Rate of climb: 1,850 ft/min (9.4 m/s)

Chapter 7
All You Need is a Hook

Canada's North was first explored earlier than might be imagined. By the late eighteenth century, there was some European cognizance of its vastness, the aboriginal people, its river systems, and its resources. To this day the North is primarily viewed as a source for fur, especially beaver; minerals, especially gold, but also copper, silver, oil, natural gas, lead, zinc and more recently diamonds. Very few white people ever saw it as a place to settle down, or even as a vacation destination. Early travelers described the North as a wasteland, a barren land, an environment of poor soil and worse climate. Snow could come in any month and frozen ground, even on a hot summer day, was never more than a few feet deep.

Unsuitable for agriculture even the few market gardens along the Alberta border struggled to turn a profit. It was too cold for cattle, and an experimental reindeer herding initiative ended badly. Where there were trees, they were spindly. Limited lumbering attempts were once and done, small-scale operations along river bottoms that faltered without government subsidies.

For holiday seekers? When the days finally warmed, the air was clouded with bugs. When the cold weather came, it was frigid, dark and interminable. The cold was delivered with an unrelenting wind.

Men did not come to this unforgiving land for a vacation. They came to the North to get something and get out. The North as a holiday destination was unthinkable, absurd, ridiculous.

For almost 300 years, starting in the mid-sixteenth century, felt hats made from beaver pelts were the European fashion of the day. The quest for beavers drove early exploration and development of the North, but barely. The first Europeans moored their ships at sea, near the mouths of rivers and enticed the indigenous people to do the traveling, trading, and bloodwork. Fur traders exchanged copper pots for pelts at astronomical profit margins. The Chipewyan Dene could argue that if you are accustomed to cooking your food by dropping hot rocks into a spruce gum sealed birch bark basket, a copper pot was worth a lot more than a stack of pelts. The Chipewyan were probably also right when they figured they were the ones swindling the white guys when they traded a pile of arctic fox pelts, useful only as diapers, for a couple of steel files. The Dene could transform steel files into knives of incredible strength or ice chisels capable of punching holes into lakes previously locked tight by mid-winter ice. White traders greedily took as much as they could get away with and sent boatloads of beaver pelts back to England for the toxic process of felt hat making. Shrewd Chipewyans revolutionized their material culture with copper pots, steel needles, guns, and gunpowder.

The aboriginals of the North also found solace in that, unlike the fate of their brethren in more temperate climes, early exploration was not followed by hordes of sod busting settlers. The only reason white traders ventured deeper into the Territories was to short circuit the trade routes to the Hudson Bay Company ships. Even then, much of the fur trading and grunt work came from Metis or First Nation people hired as middlemen.

It was collateral damage caused by the Europeans that was most horrific. Aboriginal people had no resistance to measles, small pox, and flu. Dreadful epidemics of these diseases swept across the North and wiped out the people in droves. Decimating a population also takes a terrible toll on a culture's social organization. This is not ancient history. Glenn Warner, one of Merlyn's best friends, was the founder of the Naturalist Lodge in the town of Bathurst Inlet. In 2006, he took me on a walk through the town graveyard which overlooked the Arctic Ocean. Sixty-five years earlier, a Royal Canadian Mounted Police constable sick with the flu arrived in the town by dog team. The Mountie spread the flu to the people. White crosses labeled "1941" dominate the cemetery. The settlement never fully recovered and when I visited the town less than a dozen Inuit resided there.

Intrepid European explorers ventured on exhaustive Northern expeditions with two goals: to find minerals, especially gold, and to find the mythical northwest passage - a shortcut to Asia. They had little success in meeting either goal, and the bulk of the explorers' journals dismissed the North as unsuitable for just about anything.

Journals never mention the North as a holiday destination, and usually discounted its potential as a mineral treasure trove. But these expeditions did garner press and popular interest. News focused on deprivation, hardship, and hostility. One explorer who caught the positive spirit of the land and reveled in both the natural and human history of the North was Samuel Hearne. He returned to England in 1787, but after having embraced northern life and its aboriginal people he found eighteenth century Europe insufferable and drank himself to death in five years, three years shy of his fiftieth birthday.

There were no great Indian wars in the North. Traders needed the indigenous people to trap fur-bearing animals and to transport both furs and trade goods. They needed guides to help the search for ore and a northwest passage. Unlike most of North America, the fur trade and the age of exploration was not a tip of a lance to be followed by cattle ranchers, farmers, highways, railways, and an invasion of settlers. It was different north of the sixtieth parallel. For 250 years the mantra of northern development was get as much as you could, as fast as you could, and get out. Almost no one stuck around. Only the "what was being got" changed. Beavers, whales, minerals, converted souls, musk-ox robes, fish, or just plain money. Jobs paid more in the North. In post-World War II Canada, teachers hired in the south would be transported north to teach in a tiny native hamlet. They would live in a government house. Many would bank their entire paycheck, and live off their settlement allowance, only to return "back home" in two or three years with money for a down payment on a house. Construction workers, miners, cat-skinners, dock workers, seamen, all were on a similar quest for big paychecks. Young pilots came north to accumulate pilot-in-command hours in hopes of a landing a job with a southern airlines. People talked about their time in the North as going "in" and getting "out."

The people that did stick around were the Dene, the first people. There has been criticism that the early and heady harvest of the Great Slave Lake fishery did little to help the indigenous local economy. There is truth to this allegation. Much of the money earned from fishing left the North like water from a cloud. Money-making drove everything, with little thought or regard to developing the local economy or taking care of the environment. The North historically, and in some regards even today, is thought of as a resource to be exploited. Although Euro-Canadians success with many exploits long depended on the native people's local knowledge, the North was not free from prejudice. There are plenty deplorable examples of exploitive colonialism, boarding school abuses, paternalism, and racism.

In Merlyn's time, there was no shortage of racism and derogatory comments about the Dene and Cree, but I never heard Merlyn partake in that language or behavior. Coming out of the Back Eddy Lounge one winter night past midnight, walking toward his truck, I remember Merlyn saying to me, "Did you see that Dene woman in the alley back there? She looked loaded. It's 40 below, we better take her home." and we did. There was a problem at that time with the ice road that crossed the river to the Indian Village, so we drove several miles to the steel bridge over the Hay River. It was after 2:00 am before we got home. For Merlyn, it was no more of a favor than opening a door for a lady. Not a complaint or deprecating comment was made, he was just happy to help out.

Merlyn flew caribou hauls for Indian bands, transported Metis and Dene trappers to their lines, often for barter or with dodgy cash payments far below published charter rates. He liked people regardless of color and could blend in at a party even if he was the only white guy. He would make funny, innocent, culturally naive remarks. For example, in the 1960s Merlyn would sometimes say about his Asian pilot Cedric Mah, "What I like about having a Chinaman pilot is he looks like an Eskimo, and that sometimes helps a little." Through a 2017 cultural lens it is an inappropriate comment, but when judged in the times he lived, Merlyn was accepting even progressive in his treatment and attitude toward all people.

So in the 1950s, at the cusp of the commercial fishing boom why were local Dene people absent from the commercial fishing trade? Many of the Dene did not embrace much enthusiasm for commercial fishing when hired to do its menial aspects. Sadly, racism may have played a part in the hiring policies of some of the commercial fish companies operating out of Hay River, but it is unlikely and would be ironic if that was a major factor at the Carters. Most of the crews George brought in from northern Saskatchewan were either Cree or Metis. There was little incentive to hire or to discriminate against anyone based on race. Profit was the motivation, and it made the most economic sense to hire skilled crews wherever they came from and whatever their color.

Whether from racism or economic expedience, the local indigenous people did not reap many benefits from the fishing boom. In fact, they suffered. Fish stocks were depleted in the Hay River area after a few years of aggressive commercial fishing. The local Slavey Dene subsistence fishing success declined dramatically. Eventually, the federal government banned commercial fishing in zones near Dene villages, but it has taken many years for those local fishing resources to recover.

Not all the migrants to the North got in and got out. Some people found something in the North that held them. Often it was unintentional. They fell in love with a native, got involved with an ongoing business venture or became ensnared in a government career. Some discovered, just as Samuel Hearne had in the 18th century, that the North was not the foreboding, hostile, wasteland as it had been described. From 1949 and beyond, Hay River became an epicenter for those hardy, adventurous types. Even so, Hay River's population in the last half of the twentieth century was always buoyed by transients - fishermen, construction workers, truckers, dock workers, short-time teachers. Many were looking for a quick buck. Even as late as the 1970s at Hay River restaurants, unfamiliar checkout customers were not handed a bill, rather they were asked, "Who are you with?" The assumption being that you couldn't possibly be here on your volition and your expenses must surely be being covered by your company. The transients were from all over Canada and the world. They created in the little town, a cosmopolitan culture.

Beneath this cosmopolitan culture, there was a growing cadre unique in the North to Hay River, of pioneering residents that were putting down roots. This included some who had been deepening roots since the fifties. When Merlyn first came to Hay River in 1950, he had no plan or even desire to stay, but he did. If even Merlyn and Jean were not drawn to the North by its beauty and tourist attractions, how would they entice holiday seekers? In 1962 that was the challenge, how would Merlyn and Jean create the concept of the North as a vacation destination? How would they lure tourists to the Northwest Territories?

They had some help. It took a while, but gas station companies eventually put the Mackenzie Highway on their free road maps. When they did, that thin gray line going 350 miles north from Grimshaw, Alberta, intrigued a few adventuresome tourists. Many turned back when they hit stretches of gumbo mud near Steen River, or when bullets of gravel fired by truck tires shattered their windshields, or when their cars stalled because the engine air filter was clogged with black flies. On dry days, the big eighteen-wheeler trucks would kick up a wall of dust. Tourists were blinded and learned to pull right off the highway and wait for the dust to clear. Some truckers' loads were so wide that they took up both lanes and were impossible to pass for mile after mile. Maybe some tourists just tired of an endless horizon of muskeg and black spruce. Many plans were changed and U-turns were made. But some tourists persevered. They stopped at the Northwest Territories border (the Sixtieth Parallel) and took a picture in front of the sign while fending off an onslaught of mosquitoes. They stopped at the Louise and Alexandra Falls of the Hay River. Impressive, Alexandra Falls is over 100 feet tall. (the water which had been crystal clear in 1950 became muddied by southern agriculture in the 1960s). Many of the early tourists were Americans, and they were an adventuresome lot. They were seeking danger and more relieved than scared that these waterfalls lacked guard rails or fences. Many shared the rambling, rolling stone spirit of wandering explorers. When they arrived in the town of Hay River, although awed by the sheer expanse of Great Slave Lake, they felt thwarted. Already having driven hundreds of miles, they would have been lucky to have caught even a glimpse of a moose or bear, and they likely were yet to reel in a fish. They were dusty, discouraged and hot. After noting the absolute deluge of biting insects, the second most common response of those pioneering July tourists was, "I thought it was supposed to be cold in the Arctic."

With speculation of large-scale development down the Mackenzie River and the gearing up of NTCL, Hay River was bubbling over with activity. Even so, tourist attractions and facilities were limited. Hotels were full of contracted workers, and restaurants were dark caves appointed with shag carpeting and steeped in a damp, dank smell of beer. Patrons were asked to check their knives at the door. Drinking was unrestrained and brawls not uncommon; excitement was palpable, but rowdy behavior was not the kind of excitement tourists expected. Hay River was a rough and tumble frontier town clinging to the edge of civilization, but it wasn't the wild kingdom many of the American tourists wanted. Instead of mountains, it was flat; instead of bubbling brooks it was tea colored creeks and sewage lagoons; Instead of herds of caribou; there were softball games, and the fielders all were wearing bug head-nets and drinking beer between innings. The restless tourists were searching for something more and kept on driving until they reached the end of the road where the Hay River joined the immense and majestic Great Slave Lake. They would sit transfixed and watch Carter's Cessna 180 float planes coming down the river whining, up and onto "the step" rolling onto one float and staggering airborne, hauling heavy loads to who knows where? Many of these tourists were frustrated, they were doers not watchers, and they had already invested enough time sitting in their cars. They wanted to do something, anything.

They wanted pictures, not of tugboats pushing barges heaped with oil drums, bulldozers, and building supplies, but rather photographs of shimmering trout, caribou herds, and pristine beaches. They wanted to prove to their neighbors back in Kansas City or Minneapolis that they had not been crazy to drive north into oblivion.

Inevitably, some of these tourists bumped into a big man with an endearing smile or into the big man's pint-sized bride, the plum of Meadow Lake. Merlyn and Jean were good listeners, and Merlyn knew where to find what the tourists were looking for.

It was a dilemma that spanned several years. Commercial fish flying in the sixties was still going strong and spanned the entire winter, but was always slow in the summer. Merlyn could sense that times were changing. Fish prices were flat, rising gas prices and galloping overhead costs were making interior lakes lose some of their economic advantage for commercial fishing. For the Carters, summers had always been lower key, a time to visit with friends and family back in Meadow Lake. Both Jean and Merlyn realized for Carter Air Service to be a viable business it needed to keep its planes in the sky. Its location put Hay River at a disadvantage compared to communities at the edge or soon to be at the edge of the roads. Merlyn needed to expand his market. The 180's were float equipped, and Merlyn could transport boats on the floats of these small planes. He did fly in a few sport fishermen in the 180's, but the planes were too small for a dedicated sport fishing fly-in business. If Carter Air Service was going to be able to move sport fishermen, boats, and all their gear reliably into interior lakes 200 miles away, they would need a larger float-equipped airplane. The Lockheed, although a great fish hauling bush plane, had retractable gear and could not be fitted with floats. The Beech 18 when fitted with floats makes an excellent seaplane, but Merlyn did not own floats for it. In those years, floats cost almost as much as the old planes. Merlyn was never a fan of the Beechcraft and he did not want to invest more money in that plane. He already had an ongoing lucrative summer lease for the Beech with an Edmonton firm. Buying a plane that would be primarily not a fish hauler, but a tourist transporter was a big gamble. A bigger gamble was, could the Northwest Territories ever become a tourist destination?

In most writing, the quintessential Canadian bush plane is the de Havilland Beaver. Most bush pilots would not agree with that choice. Merlyn knew he would need to build plywood tent frames and plywood tourist cabins. He never owned a Beaver as it wouldn't carry full sheets of plywood inside the cabin. When pilots lashed a stack of 4x8 sheets of plywood to the floats of a Beaver, at airspeed they would vibrate and scream like a leaf blower. A Noorduyn Norseman could vie for the title as the quintessential Canadian bush plane. It possibly is the only airplane that came equipped new with a pump-action shotgun as standard equipment. If looks and sound count it certainly wins. Fabric covered, with a big cargo door, and a radial engine that sends an eardrum-splitting roar both on take-off and at cruise, the Norseman caught the attention of the most discriminating adventure seeking tourist. Although, not a short take-off and landing aircraft the Norseman is an excellent performer and can carry a huge load. Witnesses allege that Al Loutit, a highly skilled Fort Smith bush pilot, killed wolves on frozen Northwest Territories lakes by flying low over a pack and whacking them one by one with one of the airplane's front ski.

In the spring of 1964, Merlyn purchased his first Noorduyn Norseman Mark IV. He bought FUU from Bert Berry in Uranium City, Saskatchewan. With skis to haul fish in the winter and floats to carry sport fishing passengers in summer, in 1964, the Norseman was the perfect purchase. When he bought it, the plane was sitting on floats on top of the still frozen Lake Athabasca. Merlyn had never before flown a Norseman and was skeptical about flying it without a proper check-out from another pilot. However, Berry reassured Merlyn by saying, "It's real easy. All you have to remember is three things to fly this airplane: 36" of manifold pressure on takeoff; 20 degrees of flaps and at 65 mph she'll be flying." He did everything just as Berry said he should. At 65 mph the airplane was flying. Merlyn took off in the unloaded plane right off from the ice and flew the aircraft back to Hay River, landing at the floatplane dock on the Hay River which was ice-free.

Tourists now felt tantalizing apprehension as they watched Carter's Noorduyn Norseman Mk. IV thunder down the river. The big three prong propellor turned the river to foam and pulled the plane skyward through the spray.

There is incredible fishing in the Northwest Territories, but you need a plane to get there. Tourists driving up the Mackenzie Highway were disappointed as they had expected roadside fishing spots similar to what was available along the Alaskan Highway from Dawson Creek, British Columbia, to Fairbanks. Sport anglers found themselves standing at the edge, awed by the vastness of Great Slave

Lake, but still fruitlessly seeking that fabled virgin fishing they had read about in outdoor magazines.

Merlyn's Noorduyn Norseman Mk. IV FUU
(photo courtesy of Doug Johnson)

The Carters had that something. Jean was standing on the dock ready to answer the inevitable questions. She replied to the question of whether the country to the east and north of Great Slave Lake was like Hay River truthfully with, "Oh no, it's all granite and blue water, sandy beaches and fish that have never seen a hook." That's usually all it took. Tourists would go back uptown to buy boxes of groceries at the Hudson Bay Company store, and heavier fishing line at the hardware store. Soon they would be inside the Norseman with hands over both ears heading off, hundreds of miles into the Canadian wilds.

Merlyn did know where the fish were, as he had spent several summers in the Cessna 170B scouting interior lakes for commercial fishing prospects. To the north, east, and west of Hay River, there are thousands of lakes, many of them huge. Thirteen have a greater surface area than Lake Tahoe. Depending on depth, circulation, mineral content and oxygen levels, lakes varied in their fishing potential. All the lakes had one characteristic in common, they had never been sport fished. On a single lake with a myriad of arms, such as Nonacho, anglers could have spent a lifetime scouting where the fish were, what fish were there and when was the best weeks to fish. By the early sixties, Merlyn had the lakes figured out as well as anyone. Especially the Precambrian Shield lakes in the areas east of the Slave River and northeast of Yellowknife. Merlyn sort of knew where the lunker lake trout lurked, where jackfish grew longer than a canoe paddles and where arctic grayling would rise to the sloppiest cast dry flies. Just by looking at the color of the water Merlyn could determine if a lake would hold schools of Pickerel (Walleye).

"Fish off those points," or "it will be real good in the narrows." was all the guidance necessary. Merlyn and Jean knew if they could get fishermen out to their lake camps once, and experience sport fishing in pristine water, gin clear lakes filled with gullible and big fish, the tourists would come back and they would bring their friends.

Almost without him knowing it, in the early 1960s the focus of Merlyn's flying was about to shift. What started as an effort to balance the seasons and to keep his planes aloft year around, ended up shifting most of the flying from hauling loads of plastic tubs fat with cold fish to flying loads of summer fishermen with fat wallets.

Except one thing happened. Pine Point.

Author with Nonacho Lake trout
(photo courtesy of Michael Gagnier)

Noorduyn Norseman Mk. lV
Data from Wikipedia
General characteristics
•Crew: 1
•Capacity: 10
•Length: 32 ft 4 in (9.86 m)
•Wingspan: 51 ft 6 in (15.70 m)
•Height: 10 ft 1 in (3.07 m)
•Wing area: 325 sq ft (30.2 m2)
•Empty weight: 4,240 lb (1,923 kg)
•Max takeoff weight: 7,400 lb (3,357 kg) 7,540 with floats
•Fuel capacity: 242
•Powerplant: Pratt & Whitney 600 hp (450 kW)
•Cruise speed: 150 mph; 241 km/h (130 kn) KTAS @ 10,000ft
•Stall speed: 68 mph; 109 km/h (59 kn)
•Range: 932 mi; 1,500 km (810 nmi) @ 10,000ft
•Service ceiling: 17,000 ft (5,200 m)
•Rate of climb: 591 ft/min (3.00 m/s)
•Wing loading: 22.8 lb/sq ft (111 kg/m2)
•Best rate of climb (Vy) 87 knots (161 km/h)
•Maximum - Flaps extended (Vfe) 94 knots (174 km/h)
•Maximum - operating (Vmo) 130 knots (241 km/h)/141 mph
(wheels), 138 mph (skis), 134 mph (floats)
•Stall, clean (Vs1) 65 knots (120 km/h)
•Stall, landing configuration (Vso) 59 knots (109 km/h)

Chapter 8
Pine Point a Company Town

It wasn't a new discovery. Prospectors passing by Great Slave Lake on their way to the Klondike in 1898 encountered Indians using home-made lead bullets derived from Galena deposits they dug from the ground. A Fort Resolution trapper even did some staking about 50 miles east of Hay River, but Klondike prospectors were looking for gold and silver, not lead and zinc. The Pine Point ore was forsaken, but not forgotten.

Serious interest in the mineral deposits started slowly in the 1950s but reached a fever pitch by the early 1960s. Eventually, a mine was developed and a mining town built. Cominco's Pine Point mine would become the most profitable mine in the history of Canada, pulling close to 70 million tons of ore from the ground and earning stockholders close to a half-billion dollars in profits. Most northern mines since Pine Point have gone to a fly-in, fly-out model, but Pine Point was a traditional mining town. A settlement focused on families with schools, a shopping mall, an arena and a curling rink. Even a golf course was built adjacent the mine. The Canadian government subsidized the building of a rail line from Grimshaw to Pine Point, passing through Hay River, to get the ore out. The ore was so rich that at first it was taken straight from the ground and shipped to Japan unrefined and unconcentrated. A hydroelectric dam was built on the Taltson River, 100 miles to the southeast for power. The claims were high graded in true capitalist fashion all the richest deposits were mined first.

By 1988 with the richest deposits exhausted the mine was no longer profitable, and it was shut down. Houses were sold for a pittance and moved to Hay River or down the Mackenzie Highway to northern Alberta. Schools were bulldozed, and the arena was moved to Fort Resolution. The open pits slowly started to fill with an appealing, but at the same time disturbing turquoise colored water. Pine Point was done. Had Cominco taken a more balanced approach and mined both rich and less rich deposits simultaneously, the mine could have operated profitably for a century. But they did not. They got the big bucks and got out. Despite early promises to the contrary the native people benefited little from the mine. I was a school teacher in Fort Resolution from 1977-1980 when the mine was booming, and over 1,500 people lived in the town of Pine Point. Although only 30 miles away, a gravel road connecting the communities was not built until 1972, eight years after the mine opened. Never more than a half-dozen Fort Res Chipewyan Dene worked at the mine.

The road did give residents of Fort Resolution better access to services in Pine Point and Hay River, but even that was a mixed blessing. The necessary self-sufficiency of towns off the grid results in a tighter sense of community, Fort Res lost that and gained easy access to alcohol. This was not a positive. The Fort Res K-8 Deninoo School purchased a school bus after the road came in and it was used for field trips. Ricardo Green, the principal at the time, would, palms empty, just take the bus into the Cominco's garage for routine maintenance. When it needed a new engine, and the mine bosses started to balk at paying for it, Green paused for a long moment. Cominco decided that after the hundreds of millions of dollars that they had made while carving up and desecrating many square miles of land that had been traditionally and sustainably used by aboriginal people for countless generations, they could put in a new engine in the Indians' bus. Lou Menez, the Fort Res Catholic priest, at the time, had a low opinion of the mine and cynically would say, "It has shops and schools and gyms and flowers, but it doesn't have a cemetery. It's not really a town; it will be here today; gone tomorrow." Ironically Menez was both right and wrong. It did, in fact, have a small graveyard and that's the only thing left. Everything else is gone.

The rail line, built because of the mine, had a positive effect on Hay River. It cemented the town as a transportation hub. Fuel oil, propane, steel, even buildings could be economically moved by rail to Hay River and then on to NTCL barges and pushed down the Mackenzie River. The mining fever generated by the Pine Point mine also lit a flame under the Hay River air charter companies. It ignited a mineral staking rush of mythic proportion.

In the more recent diamond staking rush north of Yellowknife, prospectors were equipped with global positioning systems and permitted to drop wooden stakes out of the windows of airplanes and helicopters. That isn't the way it worked in the fall of 1964. Hundreds of prospectors and wannabe prospectors clogged Hay River hotels and restaurants. All of them were equipped with a compass and a pile of stakes. Some of them knew how to use the compass. Most were pure speculators. Their goal was to be the first to stake everything and anything in the proximity of the Pine Point mines. It was highly competitive. One company even tried to charter every plane in Hay River just to keep their competition on the ground, unable to find a plane to charter. With the mining infrastructure now developed by Cominco, it was possible to assemble a file of claims and sell them overnight to a mining company for a quick profit.

The area around Pine Point is forested with some small sloughs, small ponds, and scattered meadows. To get the mining speculators on the ground, Merlyn needed a new plane. In northern taverns, highway construction workers talk about cat skinners that are so skilled with a bulldozer that they can "frost a cake with the blade." Aviators talk about pilots that can "land a plane on a postage stamp." To land on a "postage stamp" you have to be a good pilot and you need the right airplane. The Helio Courier is that plane, and Merlyn bought IYZ. This 4-place, STOL taildragger can fly at an astoundingly low speed of 28 mph. That means if a pilot is landing into a stiff wind he can set the plane down as easily as dropping a magazine on a coffee table. As Merlyn once said, "They don't call them Helio's for nothing." Not much separates their performance from a helicopter. Off the ground, in just a couple hundred feet and with a steep climb out, Helios are legendary bush performers. Unlike the Cessnas, they lack the wing strut. This improves visibility and reduces drag.

All this comes with some cost. Helios are challenging to fly well. They are heavy on the controls, and the big and broad tail rudder, necessary to maintain lateral control at slow flying speeds, also makes them susceptible to ground looping when landing in a crosswind. My Piper PA-20 XXO also had a big tail rudder and was short coupled, making it also prone to ground loops. A ground loop happens when a crosswind on a landing roll-out starts to push the tail perpendicular to the runway and the inertia twists the plane around on the ground in a circle. This mishap is at best, disconcerting and embarrassing. At worst, in a severe ground loop the outer wing will lift while the other wing dips and digs into the runway with a serious repair consequence. The engine on Helios is finicky too, and expensive to repair. Airframe parts are expensive and hard to find. Fuel consumption is quicker than a similar sized Cessna 180. The lack of a wing strut improves visibility for wildlife surveys. But some pilots I talked with wondered if wing/fuselage integrity is sometimes compromised without wing struts. (Allegedly, wings could fall off Helios in mid-flight, and that is never good.)

Utilizing the Helio, Merlyn and his pilots flew prospectors into impossibly tight meadows. As much of the flying was based on speculation, there was wheeling and dealing to get paid for the charters. A lot of money was floating around that crazy summer. Building sport fishing camps on interior lakes was not on Merlyn's mind.

There was a contagion aspect to the staking rush. Merlyn heard stories from his prospecting passengers, who sold claims to mine companies and made quick and exorbitant profits. Merlyn was not immune to the fever and not completely content with just doing the flying. The big money was in staking claims, if you staked the right claim.

Perhaps caught up in the fever of the staking rush, Cadillac Mines lured Merlyn into a mining business venture farther to the west. The mineral deposit of interest was located west of Fort Simpson just north of the fabled South Nahanni River. Merlyn convinced his father to help finance Cadillac's exploration and staking. Jim McAvoy and Merlyn spent a month in the summer doing the flying for this initiative. Landing the Helio on impossibly small gravel bars, and mountain meadows the duo staked many claims and dug mineral samples. But by the end of the summer, it appeared nothing of significance was found, and the company was broke. There was not money even to pay for the flying. Instead, Merlyn, George, and McAvoy were each given 100,000 shares of stock in the company. The Cadillac Mine stock at the time was not worth the paper it was printed on. But times change, and a few years later the stock was worth twenty-five cents a share. McAvoy sold his stock and walked away with $25,000 for the long-ago month's work. Merlyn and George held on to the stock, which eventually reached over five dollars a share in value. It's unclear just when George and Merlyn cashed out, but it was with the proceeds of this stock deal that Merlyn purchased his first Twin Otter MHR in 1978. It's no wonder Jean often described her Twin Otter as the "Cadillac of bush planes."

There was another prospecting story that remains a mystery. Jean's little brother Dallas Lay had become Merlyn's ace pilot. Unlike Merlyn, Lay reveled in hunting and frequently took one of the planes on forays for moose, sheep or caribou. Lay also had a keen eye for geology. A few years after the Cadillac mine staking, on one of his hunting forays, Lay discovered a vein of silver also in the Nahanni country. The vein was described as one hundred feet long and about a foot and a half wide, solid silver. Sadly in 1974 Lay was killed in a tragic Hay River car crash. Lay had died before he had a chance to do anything about all that silver and Merlyn was the only person he told. Merlyn wanted to fly into the vein with his Single Otter. He would bring a couple pick axes and a bunch of gunny sacks and "high grade" the deposit. It is not clear why he just did not want to stake it, but maybe it was already on someone else's claim, and they just did not realize the vein was there. When Merlyn heard that I was taking my daughter Lara down the Nahanni by canoe in 1997, he shared the coordinates with me and asked me to take a look. He also wanted me to find a gravel bar nearby, long enough for the Otter on wheels to land. Unfortunately, it was a few miles from the river across a boulder field of stone age proportions. After scrambling over dozens of Volkswagen sized rocks in July heat, my daughter and I needed to turn back before we reached the vein of ore. The boundaries of Nahanni National Park have expanded since then. "The lost silver vein of Dallas Lay" now lies untouched, unbeknownst to others and within the special protections of the Canadian National Park System.

The continued and extensive commercial fish flying, the staking rush, prospecting schemes, the incorporation of Carter Air Service, the flood of 1963, and three growing children kept the Carters busy. When tourists arrived in Hay River after their interests were piqued by some early marketing efforts by Merlyn and Jean, it was challenging to find the time and resources to help them.

This letter from Indianapolis lawyer, Ron Rogers, might best illustrate Merlyn's early years as a sport fishing camp operator and also his trusting nature:

At the end of my first year in law school, three of us left Atlanta on June 1, 1966, in a Ford Falcon station wagon with camping and fishing gear and a few hundred dollars and with no plans other than to see the North American continent. When we returned at the end of August, we had traveled 25,000 miles, going as far south as Mexico and as far north as just below the Arctic Circle in the Northwest Territories. We would find a camp about 9:00 every evening, Cook, eat, turn in, and be back on the road by 6:00 the next morning. Then for the next 15 hours, we would drive almost 300 miles per day, stopping to examine everything that caught our attention along the way.

By July 1, Dominion Day in Canada, we were in the Northwest Territories having traveled north from Alberta on the dirt and gravel Mackenzie highway. This was the land of the midnight sun. We stopped at a tiny fishing and prospecting town called Hay River at the southwest corner of Great Slave Lake. This big lake is as large as some of the Great Lakes and is the source of Mackenzie River which flows 1000 miles in a northwesterly direction to the Beaufort Sea. We walked along the docks looking at float planes that were used for various business purposes including mining and prospecting. The men of the town were able bodied, independent no nonsense types intent as much on surviving as much or more than getting rich. This was a working town, almost impossible to reach, certainly no destination for tourists.

One of the first men with whom I spoke was Merlyn Carter, a gentleman about five years my senior who owned the big radial engine Norseman tied up at the dock where we met. He was a pilot, a prospector and a fishing guide. He was a big man with a vice-strong grip and an engaging personality. Our trio was an anomaly in that setting. We were clearly outsiders. In a brief conversation I explained that we were American law school students on our way to the gold mining town of Yellowknife, the northernmost point on the Mackenzie Highway. Though I had known Merlyn Carter less than ten minutes, he seemed like an old friend, especially when he put his hand on my shoulder and said that he needed my help. He was in a bind. He had a customer flying in that afternoon from Indiana for a week-long fishing trip, but Carter had failed to put together anyone else for the party. This meant that he would have to spend a week alone with the man as he couldn't very well leave him alone in the wilderness for a week. It also meant he would lose a valuable week of staking mineral claims in the process. Would we be willing to pretend we were the rest of the man's fishing party and not tell him that we hadn't paid a penny for the trip? Why not? We were three young vagabonds with no schedule to keep. He tossed me the keys to his truck and told me to drive down to the town's general store and knock on the door of the proprietor, who would be asleep upstairs after celebrating the holiday. We were told to wake him up and have him open the store for us to buy a week's worth of provisions for the man from Indiana and the three of us. I was instructed by Merlyn to get some canned food items, whatever we thought was necessary, a couple cases of beer, a handful of spoons and come back and be ready to leave in about an hour. While we were gone, Merlyn loaded up the Norseman. The man from Indiana had arrived, the gas tanks had been filled, and it looked as if everything was ready for us to take off. We flew almost three hours northeast to a lake called the Nonacho. The airplane was as loud as resting your head on the side of a revving chainsaw. There was no conversation until the plane landed in the solitude of absolute wilderness. Carter taxied up to a sandy beach on an island that had recently burned and with our ears still ringing, we unloaded the aircraft. We were now hundreds of miles from our nearest neighbor.

There was a large white canvas tent, a Coleman stove, some folding chairs, a folding table, fishing gear and our provisions (including the beer). We had a big wooden boat that had been built on the island and a 6 hp motor which may have been brought in the year before and would be left on the island for another party which would come later that summer. The ice had been gone from the lake for only a couple of weeks. The water temperature was about 43 degrees. Few humans had been to this place, because it was accessible only by float plane and discovered by Merlyn Carter who had brought only a few small parties to this spot. Carter's dream was to build a fishing camp here with a few cabins and a small store for provisions.

In about 30 minutes Carter was ready to head home with the understanding that he would return one week later to take us back to civilization. Had he not kept his word or had something happened to him in the interim, we would still be there today. Before departing, he reached for his fishing rod. Carter stated that he wanted to take supper home to his wife, meaning, of course, fresh fish. I was puzzled. How long did he plan to stay to catch his supper? About 15 minutes, certainly no longer he told me. Sure, I said to myself, but not to Carter, because I am a fisherman of sorts who has sat in a boat or walked a stream all day without so much as a nibble. I was from Missouri on that one. Carter and I got into the small boat and headed out into the lake. At a random point, he stopped and told me to cast which I did with the fiberglass rod and crank reel that I brought from home. I had on my line a sizable spoon that I bought at the general store just a few hours earlier. The lure settled in the water and dropped down ten feet before I started to reel it in with a little action to make it noticeable to the fish. That is if there were actually any fish up there in the middle of nowhere. "Damn, I'm hung up Merlyn, probably a stump, this always happens to me."

"This is a glacial lake, no stumps around here." My rod bent double and I realized that this stump was moving, not acting like

stumps do down in Georgia. I fought that stump for about 10 minutes until I got it in the boat. First cast, first fish — a thirty-two pound lake trout. Merlyn casted and instantly our second fish. Carter had his meal as well as a near trophy fish and was ready to head home.

When the Norseman took off, rumbled into the distance and disappeared from sight, we were as alone and as disconnected from the civilized world as you would ever want to be. It was a good feeling. Back in the boat for another fifteen minutes, and we had supper for all. We filleted the fish, brought the cooking oil to temperature, dipped them in a light batter that Carter had given us, deep fried them on the Coleman stove, opened a few beers and had the best meal we had eaten in a very long time. We truly enjoyed our week on the island. Cell phones didn't exist. We had no radio, no TV, no newspapers, absolutely no contact with the outside world. We were in a pristine primordial place where few if any humans other than Carter and a handful of his clients had ever been before.

The lake was as clear as glass and magnified the glacial bottom down to around 30 feet. We had 24 hours of daylight. The only darkness was in the tent where we sought refuge from the light for a few hours every evening.

During that week we tried to find some bait the fish wouldn't hit. We had Mepp spinners, the Finnish Lure called a Rapala, Hoola Poppers with rubber skirts, wet flies, dry flies, spoons of all sizes and some of those big ugly plugs with treble hooks that are designed for big bass but never catch a fish and may stay in your tackle box for years untouched. We failed. Those fish were hungry and unschooled in the art of avoiding man's deception. If you got your line out any distance and brought it in at a moderate speed, you were going to get a hit. We were eventually able to see the fish following the lure toward the boat and we tried to bring in the bait so fast that they couldn't catch it. Sometimes the lure would outrun the fish but generally it did not. The average fish was 5-7 pounds with some running over twenty. The smallest fish caught during our stay weighed about a pound and a half. We kept only what we could eat and released the rest until we had caught so many we had had enough. The fish were either lake trout or great northern pike. The pike has beautiful pure white flaky meat and is one of the best eating fish I have ever encountered. The trout are slightly pink, more like salmon, but are also excellent eating. Though we had fish twice a day, we were lean from our month on the road, always hungry and never got tired of that fare.

We would eat at midnight when the sun was at its lowest point. We always kept a campfire for cooking and for warming our hands and backsides at the coolest part of the day. There was a large family, maybe a covey but they seemed more like a family, of grouse on the island that would come every evening to the campsite and watch us cook and eat. We never bothered them and they seemed to have no fear of us. There was no other wildlife on the island as far as we could tell, but bear and moose were common in the area. During that week we rested and grew fat. We were revitalized and ready to continue our trip. As the day of our departure grew near and our provisions and beer became seriously depleted, we wondered for a moment if this man we had only known for a few hours would come and pick us up. But I guess he had trusted us, and we needed to trust him.

On the seventh day and at the appointed hour the big Norseman arrived. Merlyn Carter was a man of his word, he stepped off the plane's float with a .44 magnum pistol strapped in a holster to his hip. He was a man of the North Country, a character right out of a Jack London story, honorable to a fault and larger than life. The deception was complete; the man from Indiana had a fishing trip of a lifetime and was no wiser that the three students who joined him did not pay a dime for their adventure of a lifetime.

Helio Courier
Data from Wikipedia
General characteristics
•Crew: one, pilot
•Capacity: 5 passengers
•Length: 30 ft 8 in (9.35 m)
•Wingspan: 39 ft 0 in (11.89 m)
•Height: 8 ft 10 in (2.69 m)
•Useful load: 1,320 lb (600 kg)
•Loaded weight: 3,600 lb (1,636 kg)
•Powerplant: Lycoming 295 hp (220 kW)
Performance
•Maximum speed: 148 knots (170 mph, 288 km/h)
•Range: 950 nm (1,380 miles, 1,760 km) with 120 gal. fuel
•Service ceiling: 20,500 ft (6,250 m)
•Rate of climb: 1,200 ft/min (6.1 m/s)

Chapter 9
Nonacho Fishing

It's difficult to say what was Merlyn's favorite plane. He certainly had a fondness for Cessna 180's but once he told me, "You can't fit enough stuff inside even to go camping with a 180." His favorite certainly wasn't the loud, cranky and slow to lift-off Norseman. He told Jean after his first flight in his Single Otter IOF, "I will never fly a Norseman again!" Many people who knew Merlyn might say the Single Otter was his favorite. His personality seemed to fit that big, versatile Otter. The Single Otter with its big tail and 600 hp rotary engine always drew a crowd and always got the job done. Nobody could load more in a Single Otter than Merlyn. In fact, Merlyn often carried more in his Single Otters than some other companies dared carry in a Twin Otter. Long after the Carters purchased the Twin Otter Merlyn often still flew the Single and left the Twin to his son Dean or hired pilots. Maybe that was only because no one else could fly a Single Otter with the aplomb of Merlyn. Or was his favorite the Twin Otter? The plane Jean called the Cadillac of bush planes. Make no mistake; Merlyn liked the Twin Otter. I remember the first time he landed on Nonacho Lake in his freshly painted Twin Otter MHR. With the twin propellors and the ability to reverse the thrust, he spun the plane and backed it up on the beach.

I remember him climbing out on the float with a boyish grin on his face. I said to him, "You could never do that trick with the Single Otter."

"Yeah, no water rudders." he said, with again the kind of smile on his face that a boy has when he kisses a girl for the first time, and she kisses him back.

There was a drawback of not having those pesky water-rudders. If you lose an engine on a taxiing float-equipped Twin Otter, it is almost impossible to steer and dock the aircraft. Merlyn was able to do just that on Four-Mile Lake while another pilot who first attempted it just kept going around in circles. Merlyn had a legendary ability to control crippled aircraft both on the ground and in the air. He once managed to take-off with the Anson on one engine, and more than once took off and flew loaded airplanes crippled by a flat tire.

But I don't think it was the 180, or the Otter or even the Twin Otter that was Merlyn's favorite plane. I believe it was the Lockheed 10A Electra HTV. If things had worked out differently, he might have kept the Lockheed forever. Maybe he would have had it restored and flown it to airshows with Jean. It was the plane Amelia Earhart had made famous and the plane Joe McBryan called a marvel of its time. Merlyn flew the Lockheed with the grace of a dancer, an Argentinian tango dancer.

Merlyn's Twin Otter taking off from Hay River after switching over from wheels (courtesy of Carter family album)

But things did not turn out that way. Here's what happened.

Merlyn was hired to haul fuel 80 miles north of Hay River for a seismic company searching for oil fields. The company had told him that the muskeg was frozen hard and that there was a suitable place to land at the camp. It was early autumn, and Merlyn wasn't convinced the weather had been cold enough to drive the frost down into the muskeg, so he only loaded two drums for the first flight.

When he landed, and as he slowed, the weight from the loss of lift settled in, and the Lockheed broke through the frozen crust of the muskeg. The main gear dug in and stopped. The plane's forward momentum lifted the tail high into the air. Merlyn remembered the plane teetering on its nose. Just as he thought the plane might fall back onto its tailwheel, it completed its flip and fell over onto its back.

Hanging upside down in the plane, Merlyn struggled to free himself from his seatbelt. It was the old style of seatbelt that is held tight by friction. Upside down the weight of his body pressing on the seatbelt made it difficult to unbuckle. As he struggled, he heard two disconcerting noises. The crash had tumbled his cargo, and he could hear the diesel fuel trickling out from one of the barrels, and he saw it pooling in the upended cabin roof. The second sound was an electrical short, rhythmically firing a spark. This was not a good combination. First, he turned off the master switch to kill the electrics, and then with a super human effort, he pushed himself with one hand against the ceiling to back off the weight on his seat belt so he could release it. He landed on the plane's fuel soaked ceiling with a thud, and in the shape of a snail.

Before he could extract himself from the cabin, he needed to fiddle with the door latch. Because of the flip, everything was upside down and backward. He finally got out, struggled to his feet and walked around the plane. Merlyn thought it didn't look too bad, only two bent props. Already thinking of salvage, he presumed with some help he could build a tripod of spruce logs, hook on to the tailwheel and pull the plane back upright and flip it back over. Then he could fly in two new props to replace the bent ones. Once the ground was frozen hard, he would be able to fly the plane out. Still drenched with diesel fuel, he started walking towards the camp about a half-mile away to use the HF (High Frequency) radio and call home. When he got 200 feet away from the plane, there was a HARUMPP that shook the ground, and a brilliant flash lit the autumn sky. He turned in time to see his cherished Electra bursting into flames with enough force to blast the tail section twenty feet in the air and to litter the ground with burning debris.

A few hours later one of his pilots came in with an empty Cessna, landed gingerly, and picked Merlyn up, returning him to Hay River. Merlyn went back to the crash site the following day, only to see the plane still burning "I can't believe something made out of metal could burn that long," he said with a chuckle, and then, as he always did after crashing an airplane, Merlyn went on a three-day bender. And he never again fastened his seat belt.

The demise of the Lockheed in 1972 was a turning point in Carter Air Service. Winter commercial fishing in Hay River had continued a decade long decline. Merlyn believed the establishment of the government owned and controlled Freshwater Fish Marketing Corporation would seal commercial fishing's dismal fate, and he was right. In his lifetime, fishing harvests would never again approach the levels of the 1950s.

Fishing had not ended, but fishermen coped with greater and greater overhead while the price of fish stagnated. The old timers still plied the lake in summer by boat and winter by Bombardier, but the younger generation questioned the hard toil for such meager returns.

Merlyn was still flying fish, but increasingly the fish were being caught on hooks, not in nets.

The Carters could not count on the trickle of tourists passing through Hay River as their only source of sport fishing clients. They needed to establish the Northwest Territories as a world class destination for fishermen to visit. They did this by renting booths at fishing sport's shows in major Canadian and American cities. Merlyn and Jean would load up one of their 180's and "do the show." Never half-way, Merlyn and Jean did the shows a little differently than anyone else. Before heading to Edmonton, Kansas City, or Minneapolis, they would ask one of their commercial fishing friends to save them a big lake trout. Along with some folksy brochures and a scrapbook of photos, they would bring along a real 50-pound frozen lake trout. Lodges from Ontario and Alaska brought fancy stuffed and mounted fish to the shows, but the Carter Air Service booth drew the crowds. Glistening on a bed of ice, everyone wanted to see the "real" fish. When Merlyn pulled out his guitar and started singing, and "the plum of Meadow Lake" teased potential clients with witty remarks and a million-dollar smile, people signed up. The right type of people signed up: risk takers. Most Americans had never even heard of the Northwest Territories. Alaska and Ontario were the safe bets. Merlyn and Jean hooked the adventurers.

Lara Kesselring at Nonacho Lake
(courtesy of Carter family)

There were a few other booths at the shows promoting fishing in the Territories. Their lodges were different. Tourist operators bought played-out mines and abandoned trading posts on Great Slave Lake and Great Bear Lake for a pittance. Some of these sites had airstrips, power plants, plumbing, and hotel-like accommodations. Cheap barge transport could resupply these sites in the summer. Developers could transform the old buildings into functional fishing lodges with dining rooms, bars, showers, and docks. This was not the way the Carters developed their business. Merlyn brought fishermen to virtually unfished, pristine lakes in the deep interior of the Northwest

Territories, far from any road and off the circuit of Great Bear Lake and Great Slave Lake that were reachable by summer barges and winter ice roads. Every outboard motor, every sheet of plywood, every gallon of gas, every bottle of beer had to be flown to Carters' camps on the interior lakes by float or ski plane. The lodges on the big lakes had great fishing, but the area around the lodges had been fished hard already, and guides often had to take clients many miles to reach the type of great fishing Merlyn and Jean could promise right from their beach. "Big" is also a relative term. Nonacho and Point lakes are huge bodies of water and if located in the continental United States would be among the largest lakes in the country. Nonacho is sixty miles long, and Point Lake is even bigger.

Merlyn knew where the big fish were, and he established camps on over a half-dozen lakes, including Point Lake, McKay Lake, Gagnon Lake and Nonacho Lake. In the first few years, the Carter outfitting was basic. Merlyn hired Cree fishermen to make wooden boats in their West Channel backyards. For Merlyn to fly the boats to the lake, it was necessary to saw them in half and tie them to the floats. The boats were reassembled at the lake and made watertight, or close to watertight, with tar. These boats were affectionately called "Tar Babies." A small outboard motor, a 55-gallon barrel of fuel, maybe a wobble pump, several cases of beer, a couple of tents and camp gear, a venerable Coleman white gas stove and a rusty WWII Enfield .303 to ward off bears is what clients got. The fisherman didn't complain about roughing it. They were catching more trout and bigger trout in an afternoon than they would have caught all summer in northern Ontario. Caribou were walking through camp, wolves were howling from the eskers, tooth-tingling drinking water was within an arm's reach of the boat. It was exactly the kind of adventure they wanted, and they snapped photos to brag and share with their neighbors back home. In the 1960s the climate was colder, and the big lakes never warmed up. Fishermen could troll on the surface all summer with lures the size of mixing spoons and expect to catch fish. For most fishermen, the most esteemed fish was the lake trout. The size of Nonacho lake trout was astounding. Fishermen commonly caught 20-pounders, 30-pounders were not unusual, and sometimes 40 and even 50-pound lake trout were caught. The flesh of a lake trout caught in northern waters is orange or sometimes pinkish. Freshly caught and fried over a driftwood fire or on the Coleman stove resulted in sublime eating. Some of the lakes also contained large populations of great northern pike. These too were huge, sometimes larger in length than world record pike, but smaller in girth and weight. Their slender shape earned them the nickname "Snake". To the locals who held them in disdain, they were also called jackfish or jacks. Fishermen often caught jacks while trolling for trout, but sometimes anglers would zero in on the jacks by fishing in the shallows of bays. Trout take the bait and dive, pulling as if an anchor was attached to the line. Jacks often slammed the lure at the surface and were said to hit it like a crocodile. Many of the lakes also contained arctic grayling - a dark, almost purple fish with a sail for a dorsal fin. Grayling have a small mouth and are best caught with light tackle, often jumping clear out of the water like a rainbow trout. Modest in size and usually thought of as a river fish, it might be of interest to note that grayling can be quite common in northern lakes. Also, lake trout often thought of as a lake fish can be quite common in northern rivers.

Merlyn's camps on several lakes did not disappoint, but over the years he narrowed his focus. Barren ground grizzlies kept tearing up his camps on Point Lake. McKay Lake was also on the tundra and prone to high winds and rough water. Gagnon was an excellent lake, but gradually more and more, Merlyn settled on Nonacho Lake as his favorite. One of the largest interior lakes, Nonacho is spidery in shape with many protected arms and bays. The current of the Taltson River flows through the lake for 29 miles and perhaps the nutrients created by that flow boost both the population and the size of the fish. Back in the 1950s when he was checking interior lakes for their commercial fishing potential, Merlyn had remembered netting a Nonacho lake trout that exceeded sixty pounds.

Nonacho Lake was fished commercially by Carter Fisheries, but only briefly. The flight to Hay River or even form the east arm of Great Slave Lake was too long to be profitable. The big money fish were jumbo whitefish and although Nonacho held many jumbos, the

rance of large lake trout, and their twisting nature when
ed to tangled and destroyed nets.

Merlyn discovered a perfect site on Nonacho for a sport fishing
camp--a protected island with a southwest exposure. It had a broad
sandy beach and was perched right at a narrows in the lake. Nonacho
Lake is one of the most beautiful lakes in the Territories. The lake is
rimmed with ribbons of sandy eskers, open meadows of tundra and
glades of jackpine. It was a favorite wintering ground for the Beverly
Caribou Herd, estimated in the 1970s to include over 425,000
animals. In pre-contact times, it was also a favored winter camping
area for the Chipewyan Dene people.

What started as a couple of canvas tents on the beach evolved
over the years into a little settlement. The first step-up was a wood
framed wall tent. With local, peeled spruce utilized as floor joists, a
floor of plywood was laid down, and then short plywood walls were
erected, and a plywood door installed. The rest of the structure was
white canvas. Canvas wall tents are bright and comfortable. The
canvas breathes so they are cool on hot days, and with a little tin
wood stove, they heat up quickly on frosty mornings. Soon the wall
tents were relegated to storage or to accommodate camp workers,
and 16-foot by 16-foot plywood cabins were built. Coleman stoves
were replaced with propane units. An ice house was built, and ice cut
from the lake in the spring lasted the entire summer. The ice kept the
beer cold and the fish fresh until a plane came in. A wheelbarrow to
haul gear and baggage was soon replaced with an old crank-starting
tractor. Merlyn flew out the tractor in early spring in the the ski
equipped Otter. Even with the tractor's wheels removed, the tractor
cab stock half-way out of the cargo hatch of the plane. Pianos, boats,
generators, tractors, Merlyn loved the challenge of transporting
oversized, overweight external loads. A little store was built in the
summer of 1977, and a satellite phone ultimately supplanted the
raspy static of the High Frequency (HF) radio. Merlyn was selling
groceries to fishermen almost like his father had done in the West
Channel twenty-five years earlier. About the same time, the first
diesel powerplant was flown in, again hanging half out of the Otter
for over 200 miles. Dug into the slope of the esker behind his camp,

the incessant throb of its churning cylinders made me, for the first time, regret building my cabin only a mile from the Carter camp. With the purchase of the Twin Otter in 1978, the flood gates of cargo opened wider. Running water, showers, additions to the cabins, an air-conditioner for the store, washers, dryers, barbecues, a Ford loader/backhoe, an ice machine, satellite TV and more.

As much as Merlyn enjoyed sitting in front of the air-conditioner on a hot July day, he would come over to my cabin on occasion. There we would sit with our backs against my log house and drink a few warm beers. Merlyn would open the beer bottles with a pair of pliers and talk about the days when the camp was simpler. He did not say much. Mostly he sat quietly. Silence can be a mask. It would be a mistake to believe Merlyn was not a deep thinker. It would be a mistake to assume he had a happy-go-lucky, nothing fazes me, persona. It is even wrong to believe that he never questioned his career as a bush pilot. True enough, he was quiet about his feelings. Merlyn was always at the center of a party and could fly 24/7 with nary a complaint. Still, he often sat in silent reverie. He hid a restless side that only rarely leaked out. He spent a lifetime with no brothers with whom to confide, a laconic and austere father and a mother he felt he needed to care for and protect from the day-to-day challenges of life. Even with Jean, Merlyn was more of a listener than a talker. His wife who could juggle a dozen tasks with utmost efficiency and could keep track of the comings and goings of dozens of tourists directed the day-to-day business operations. Inevitably, there were moments when she would feel overwhelmed and would collapse into her husband's arms and requiring a listening ear. Merlyn had his demons, his moments of self-doubt and deep thoughts, but he was stoic.

He could be blue and experience a deeper melancholy especially when winter flying was quiet and the days were short. In his Vale Island home on a hard chair next to the woodstove he would sometimes sit motionless with vacant eyes. Once, on the eve of his granddaughter Tiffany's wedding we were sitting together in his truck. I played a bush pilot song by Dave Hadfield on his CD player.

Merlyn turned to me with tears in his eyes and said, "I don't know Rob, I don't know what I have done with my life, I just don't know."

I could remember him two decades earlier sitting against my cabin wall at Nonacho and saying much the same thing. Already a fabled northern aviator, slowly drinking a beer, he capriciously mused about quitting it all and driving truck. But he would quickly recover with, "Well Rob, you've been all over the world, and you keep coming back up here. We got lots of grub, and I can tell you there are a lot of people who don't." Then he bottled-up his earlier thoughts, roused himself up, making sure he still had his little notebook and pocket of tools and headed back to his camp to tinker with the tractor or generator or the ice machine.

Merlyn may have given the impression that he never second guessed anything such as, making a forced landing at the end of the runway because his Otter was overloaded; frying his Nonacho camp powerplant when he replaced a fuse with a piece of copper pipe, or bashing in the cab of a brand-new truck when he tried to load a deep-freeze refrigerator by resting it on the gate, revving the truck backward and slamming on the brakes.

He seemed unruffled when his pilot nephew, Stu Poirier, put CZP through the ice at Rocher River and unperturbed by the parade of blown engines, leaky floats, outrageous insurance premiums, daunting and ridiculous new regulations imposed by Transport Canada or the Government of the Northwest Territories.

Merlyn gave the impression he never second guessed anything. In reality, I think he second guessed everything, especially everything of substance. Loyal to his many friends and relatives he would visit, listen to their stories and empathize with their struggles. His daughter, Kandee, remarked on how her dad brought out the best in others. His storytelling calmed an audience and created a mood that all was right in the world. It did not matter much if he was in a smoke-filled Cree cabin at the West Channel or with high rolling clients at his camp. Merlyn found time for everyone, his cousin Elaine, was another of his best friends. He never visited Meadow Lake without stopping for a visit with her.

Merlyn never bothered to read a novel and truth be knc reading skills would likely have prevented even that possibili˄ ˄ it would be wrong to assume Merlyn did not nurture deep thoughts and wrong to assume he lacked wisdom. The greatness of a man can, in some ways, be measured by his uncertainties. It is a fool that questions nothing and lives an unexamined life. Merlyn had his doubts, his deep thoughts and his contradictions. I believe he was one of the wisest men I ever knew. He just did not wear those thoughts on his sleeve.

It was with mixed feelings that Merlyn kept expanding his camp. He liked the ice machine because for a man who dealt ten months a year with the vagaries of ice the last thing he wanted to do was fly ice in the summer. Filling the ice house in the spring with blocks of ice cut from the lake with a chainsaw was an arduous chore. What was his solution to ice in 1975 when my first wife and I lived a year in a log cabin at Nonacho? He bought a gas-powered water pump and had us fill the ice house by pumping four inches of water from the lake into the house every week. The resulting giant block of ice 16 x 16 x 8 feet lasted two years. The ice house was never filled again. After that, it was an electric ice machine. Merlyn did not mind the throb of a diesel generator. It reassured him. He loved the wilderness, and he loved the conquering of it even more.

Merlyn liked his tourist clients to have clean boats. It was one thing he insisted. Customers wanted big fish but also appreciated better equipment and more comfortable lodging. Tourists wanted seaworthy boats equipped with higher and higher horsepower. The days of tar babies with 6-hp Evinrudes were short lived. Even his fleet of 14-foot aluminum boats were undersized for a lake the size of Nonacho. As years passed the lake ice started melting earlier in the spring and freezing later in the fall. The method of fishing changed, by mid-summer the big trout began seeking out deep holes. For the first couple decades, there was always ice on the lake until the last week in June and sometimes into July. In the sixties and seventies, Merlyn would never book a tourist trip earlier than the last week of June and sometimes even that group would be dealing with ice pans. By the nineties sometimes Nonacho was wide open in May.

Freeze-up was correspondingly later going from mid-October to mid-November. This longer ice-free season meant more time for the sun to warm the waters and mid-summer surface water temperatures increased. In the early years, trout could be caught at the surface all summer long. Twenty years later by mid-July the trout had sought deeper, colder water. It did not cause fishing to deteriorate, just change. Fishermen needed to fish the deep holes in mid-season, and this sometimes meant more travel on the lake. Merlyn needed 16-foot boats.

How could he get 16-foot Lund boats to Nonacho? At the time Merlyn still owned IOF and he could have strapped them, one at a time, to the floats of that Single Otter. But Merlyn loved the challenge of carrying external loads and I knew it bugged him that external loads were forbidden on a Twin Otter. The reason it is illegal to carry an external load on a Twin Otter is because if a pilot loses an engine on the side of the plane, the boat is tied on, there won't be enough rudder or rudder trim to keep the airplane flying straight. A failure of that engine would result in a deadly crash. But Merlyn had it figured out; he would tie a boat to each side of the plane to keep the load even. The load would be balanced regardless if he lost an engine. Merlyn never tested this idea but it made sense.

So, this is what he did: "I will tell you how we are going to do this Myles. We are not going to tie the boats on during the day. People are around, other pilots, even Transport Canada sometimes comes down. But they're government; they won't work after 5 o'clock. We will go down to the plane at 10 o'clock tonight." It was June and daylight all the time. So that's what he did. Merlyn and Myles spent a couple of hours in the middle the night, tying the boats on to the floats, one on each side. They went home, took a nap, and came back at 4:00 am and flew the boats out to Nonacho Lake. He did that mission four times and in that way got all eight boats to his camp. One consensus from all the old pilots of the North: The Territories would have never been opened up without aircraft hauling external loads and overloads. Merlyn epitomized that consensus and viewed regulations as similar to bad weather — something you had to fly around, beneath or through.

The sheer quantity of fish caught by tourists at Nonacho was astounding. Even people who didn't know a spoon from a spinner could catch a trophy fish. Skilled fishermen caught boatloads of trout. The concept of "catch and release" fishing did not gain traction at Nonacho until the end of the millennium, but tourists from the outside usually stuck to the legal limit. In the 1970s There were a couple of groups from Silver Bay, Minnesota, that returned every year. The Carters particularly enjoyed hosting them. They knew fishing, and they were self-sufficient, often leaving the Carter's outboard motors running smoother when they left than when they arrived. It was their goal that every person take home their limit, without a fish under twenty pounds. In the days before the ice house, they devised ways to keep caught fish fresh by digging pits into the permafrost or building pens in shallow water. These guys were hard-core Minnesota fishermen and knew how to fish and how to live in the bush. They returned every year often bringing more and more friends and family members with them.

One of Merlyn's dearest friendships began in 1972 when NHL and WHL star hockey player Jim Harrison came fishing at Nonacho. Harrison who once scored 10 points in a single game described Merlyn as his "best friend" and told me, "I cannot count the bottles of rum we drank together." The summer was a time for professional hockey players to relax and Harrison brought many players to Merlyn's sport fishing camps, including such all-stars as Bobby Orr. Once when Orr was fishing at Nonacho Lake word got around through Merlyn's best friend, Don Boxer in Fort Smith that Bobby Orr was fishing at Nonacho and soon four float planes were circling his camp. Everyone wanted to meet Bobby Orr.

Harrison eventually built his own cabin on Nonacho Lake. Merlyn frequently went to Chicago to watch games when Harrison was playing for the Blackhawks and bought season tickets when Harrison was traded to the Edmonton Oilers. Harrison invited Merlyn to after-game player's parties. These were rowdy affairs, and Harrison shakes his head when he remembers Merlyn walking in the door. "He would be wearing his northern parka, front shirt pocket bulging with a little notebook, a pair of pliers and a roll of tape.

Rubber boots were on his feet." As always, Merlyn was at ease. Other guests initially gave him awkward looks, wondering who this bushman was. But an hour later not only was he accepted, Merlyn was the center of attention. Singing songs, with women twenty years younger on both knees, doing his egg magic trick, telling stories, and always wearing a smile that warmed peoples' hearts and prompted them to drop all pretenses and spontaneously snuggle up to Merlyn. In light of the nature of his demise, it is somewhat ironic that friends frequently described Merlyn as a big teddy bear. Whereas his father was reserved, laconic and formal, Merlyn was affable, open and welcoming. Even in the late 1950s, when Merlyn was in his early twenties, employees with a beef about Carter Fisheries wanted to talk with Merlyn, not George. Merlyn had the rare ability to earn people's trust almost immediately and without effort. That never changed.

Nonacho Lake also became a local destination especially for private business parties and family celebrations. Locals from Hay River were less likely to adhere to legal fish limits. I can remember loading 32-gallon Rubbermaid garbage cans into Merlyn's Otters that were filled to the brim with trout fillets.

Some came as much for the drinking as for the fishing. A few had the balls of Hercules but the brain of Donald Duck. The party often began when they stepped into the airplane from Hay River dock. The Single Otter was known for its short take-off and landing capabilities (STOL) and its ability to carry a big load, but it was not known for its speed. Depending on the wind it could be a two-and-a-half-hour ride from Hay River to Nonacho. After drinking several beers that liquid has got to go somewhere, and the vibrating Otter had a way of speeding its journey to bladders. If there was a woman in the plane in distress, Merlyn would drop down, land on a lake, and pull up on a sandy beach. For men? "That's what cowboy boots are for." Harrison does, however, remember a flight when someone needed a mixer for Crown Royal and Merlyn willingly dropped down on to a nameless lake to fill an empty barf bag with water.

Locals also tended to overestimate their navigation ability in the fishing boats. They would zoom away on a fishing excursion from Nonacho's beach with outboard motor roaring and little more than fishing tackle and a couple of cases of beer in the boat. Several hours later the beer and the outboard's gas would be consumed, and they would be stranded drifting aimlessly under the midnight twilight. If Merlyn was in camp he would throw some cans of mixed gas in the cabin of the Otter, bring along his teenage son, Dean, and find the lost fishermen from upstairs. Although Nonacho Lake is big, the guys were never too far from camp and it is easy to spot a floating boat from the air. Merlyn would land next to the boat and Dean, who had an uncanny knowledge of Nonacho's many islands and bays, would bring the crew back to camp. Merlyn never charged for these rescues. The Carters never charged for anything once tourists reached the camp. Drunk fishermen would tear the lower unit off a kicker when they hit a reef, or break a cabin window, or bust a chair, and Merlyn would just shrug his shoulders. As a result, tourists would come back time and time again. They would send the Carters Christmas presents, and feel as if they belonged, not just as customers, but almost as extended family.

One time, there was not a plane in camp when a group Hay River guests became lost. When the fishermen had headed out they were careful that all the beer bottles were full of beer, but neglected to fill their gas tanks with fuel. After catching a couple stringers of fish, they were unable to find camp, and wove their way through the island studded lake until their outboard motor ran out of gas. Cold, and knowing that Merlyn was not in camp and that there would be no aerial search, they decided the only way Dean would find them is if they made a signal. They decided to light an island on fire. Even this attempt failed when the island failed to ignite. Early the next morning, Dean found them by boat. Two days later flying home to Hay River in the Otter they saw "their" island now ablaze in flame. The burning island was cause for a lot of laughter; the cold night was forgotten and once again the Nonacho fishing trip was a memorable and fun experience.

To be fair, navigation in the far north can be tricky, especially on a big spidery lake like Nonacho. Instead of up and over, the sun moves around you in a circle. This messes with a boater's internal sense of direction. A compass can be of little help because being so close to the pole the needle is sluggish and also easily distracted by the vast mineral deposits of the Precambrian Shield. There are no big landmarks on the horizons, no Mt. Raniers, no rivers, just a myriad of islands that all look similar. More than once I have been in a cold sweat on that lake, looking fruitlessly for something familiar to help guide me home. What I will never understand is the ability of the native people to navigate northern wilderness, not by compass or map, but by memory. When I lived in Fort Resolution, I went out hunting with members of the community. Their ability to find their way in the wilderness was uncanny. I remember accompanying one Chipewyan elder who must have been close to sixty find his way to an old cabin site he had not visited since he was a kid, a half-century earlier. He could still remember and remark about tiny creeks even odd shaped rocks. Merlyn had that same ability from the air. Over huge spans of topography, his paper maps would remain stashed in the door pocket, and he would find his way to the most remote locations by memory alone.

I remember accompanying a few Chipewyan hunters on a moose hunt by canoe up the Slave River. Before the trip I showed up with all my topographic maps and asked where we were going to camp. They looked at each other partially incredulously and partially humorously, "We will camp where we shoot the moose." They laughed as I realized it would not have made much sense to drag a 1500-pound carcass to a pre-arranged campsite. Merlyn had the same kind of matter-of-fact good sense.

The success of Carters' tourist camps did not go unnoticed. Other Hay Riverites, maybe as much for the fun as profit, started building camps on interior lakes. Tourist camps were built on Rutledge, Thuban, O'Connor and Dogface Lakes. Carter Air Service also flew passengers for Brabant Lodge on Great Slave Lake. Flying building supplies, provisions and passengers these camps brought substantial revenue to Carter Air Service. Canadian charter companies charge by the mile both to and from a destination. Merlyn was able to help these camp owners by splitting their loads with Nonacho runs. If there was a load of passengers going to Rutledge, Merlyn could continue to Nonacho and bring back his guests out on the same flight. That way Bob McMeekin, the owner of Rutledge would only have to pay a one-way (or half) the charter. This willingness to work cooperatively with other lodge owners earned the Carters extreme loyalty in Hay River. Charter Carter was everyone's first choice. It also meant there were some interesting passengers on the flights to the other camps. One lodge was known for high stakes gambling, another for prostitution. The Carter camp was known for fishing.

After figuring in the costs of camp building, maintenance, and the southern sport show advertising, the Carter camps themselves did not make a lot of money. What they did do, and what the other camps did, was keep Merlyn's airplanes in the sky all summer, and that is where the money was.

Merlyn may have been at his funniest when he settled disputes. It was 1975 in early September of my 23rd year. My wife and I had finished building our log cabin at Nonacho - a seven-week task. Earlier in the day while we were chinking the cracks between cabin logs with moss, we heard the rumbling Single Otter fly into the Carter camp. It had been a couple of weeks since any tourists had visited the camp so we thought this might just be the family. Jean was a great cook, and we knew it. We had been living on a diet of fish. A little before suppertime we decided to paddle the mile over to the camp. The glacial esker that connected our cabin almost to Carter's island was, in fall, crimsoned with cranberries and Jean, unbeknownst to us at the time, would always come out for a few days of September berry picking. This time, the Carters had brought along some relatives from northern Saskatchewan. We were invited to join the family dinner table.

Between Jean and Merlyn there were many relatives. When one of your relatives owns a float plane and a sport fishing lodge, you become close to that relative. Merlyn and Jean were gracious and tried to host many of their relatives out at the camp every year. It would be impossible to imagine more interesting guests. Old commercial fishermen would set nets; there would be yodeling guitar players, stories from Jehovah Witness missionaries and a guy that competed in the chuck wagon race at the Calgary Stampede.

Everyone brought something to the party and everyone had a story to share. The two visitors on that September day were somehow related, but I knew not how.

In a 16 x 16 foot cabin with a little propane stove, Jean could put together quite a supper, including marinated caribou steaks, perogies fried in butter, three kinds of vegetables, fresh baked biscuits and pie for dessert. I did not know at the time that Jean had honed these cooking skills under her mother's tutelage many years ago in a smaller kitchen than the kitchen counter at a Nonacho tourist cabin.

After dinner, before Merlyn had a chance to pull out his guitar, a discussion ensued on how my wife Bonnie and I had decided to build a log cabin in the middle of nowhere and spend the winter in the bush. I shared that we had become excited a year ago when we paddled a canoe, 1600 miles, from Fort Nelson, British Columbia to Tuktoyaktuk on the Arctic coast. Recounting our route perplexed the family relatives.

"How could you get from Great Slave Lake to the Arctic coast in a canoe?" they asked.

"We paddled down the Mackenzie River" we responded.

"Down the Mackenzie River??? That's not possible. The river flows south from the Arctic Ocean to the Great Slave Lake." They were emphatic.

The discussion became a bit of an argument. Just a year ago I had graduated from the University of Colorado with a bachelor's degree in Geography. I knew which way the river flowed. Exasperated, I finally said, "If the Mackenzie flowed from the ocean to the lake, the lake would be saltwater."

They were unconvinced and looked at us incredulously.

Frustrated, I turned to Merlyn, "Tell these people which way the Mackenzie River flows."

Merlyn paused for a moment, licked his lips as he was apt to do prior to a story and said, "Well Rob, look at the map. The Arctic Ocean is up there on the top, Great Slave Lake is on the bottom, of course, the river is going to flow from top to bottom."

His family guests were vindicated. Merlyn laughed so hard his whole body shook. I gave up the argument.

If there was one lesson I learned from Merlyn it was if you have to choose between being right and being kind, it is better to choose kindness.

While I was working at his camp on Nonacho, bears would occasionally wander into camp attracted by the gut bucket or just human activity. I was always able to roust them out of camp by chasing them and yelling and with a parting salvo of bird shot from my ten-gauge sawed-off shotgun. Never a lethal blast, but still the raspberry on the hide of a bear from two ounces of number 8 shot delivered a powerful message. I never had a bear return to camp a second time. I remember Merlyn telling me, "That's real good, Rob I don't like killing those bears."

A few years later when a trigger-happy, crusty, Leonard Anderman was watching the camp, he would kill every bear that wandered remotely close to camp, and I overheard Merlyn saying to him, "That's real good Leonard, kill those buggers." In light of later events maybe Anderman had the right idea, but the point was if it wasn't a big deal, Merlyn went with whichever way the wind was blowing.

The first time I flew with Merlyn in the Single Otter, I was in the right seat. It was my first float plane ride ever. I was digging around looking for the seatbelt, and I asked Merlyn for help. He pulled the belt out from the crack and said: "Yeah, you can use that if you want, I don't, but you can." He did what he wanted, and he never judged others for doing what they wanted.

Merlyn dreamed big. He had a chance to buy a DC-6 real cheap. A DC-6 is a four propellor airliner. The engines on this old bird were timed out, and there is no demand for a DC-6. Here was his plan. Fill the DC-6 with fuel barrels and fly it to Nonacho in April, land on the ice, spin it around, and park it permanently on the beach. He figured he could bring in three seasons worth — 7,500 US gallons of fuel — all in one load.

Thinking the plan over he had second thoughts. Although Merlyn has had minimal problems with people stealing anything from the lodge, he was concerned about creating such an immense cache of fuel. More than that it was a little like buying a Halloween costume the day after Halloween. The 50% savings is appealing, but spending money for something too far ahead is difficult. Merlyn was never great at postponing his gratification.

April was the best month to haul building supplies to Nonacho Lake (courtesy of Carter family album)

Jean on the tail elevator, Merlyn on the ice. Building their camp at Nonacho (courtesy of Carter family album)

Eugene McKay with Dean Carter (by cargo door) and possibly Merlyn and Jean's granddaughter Tiffany in the toboggan as they load up in Hay River (photo unknown origin)

Merlyn liked the challenge of impossible external loads. Here's a 14-foot Lund fishing boat tied to the pontoon of Cessna 180 TWN (photo courtesy of Carter family album)

Party time at Nonacho Lake. Merlyn was a talented country singer and guitar player, and a magician of some note. (photo by author)

De Havilland Canada DHC-6 Twin Otter Series 100
Data from Wikipedia
General characteristics
Flight deck crew 1–2
Seating19
Length 51 ft 9 in (15.77 m)
Wingspan 65 ft 0 in (19.8 m)
Wing area 420 sq ft (39 m^2)
Empty weight 5,850l lb(2,653 kg)
Maximum Weight 11,566 lb (5,246 kg)
Height 19 ft 4 in (5.9 m)
Performance
Maximum speed 160 knots (297 km/h at cruise altitude)
Cruise speed 150 knots (278 km/h at cruise altitude)
Stall speed 58 knots (107 km/h at cruise altitude)
Range (Max fuel, no payload) 771 nmi (1,427 km)
Maximum fuel capacity 382 US gal (1,447 L)
Service ceiling 25,000 ft (7,620 m)
Powerplants (×2) Pratt & Whitney PT6A-20550 shp each
Rate of climb 1,600 ft/min (8.1 m/s)

Chapter 10
Caribou Hauling

In 1980, I flew with Merlyn in the Single Otter CZP to pick up some First Nation Slavey who were moose hunting on Buffalo Lake about sixty miles southeast of Hay River. As we were leaving the house, Jean said to Merlyn, "Make sure you get paid, those guys haven't even paid for their ride out yet."

Merlyn replied, "Yup, got it."

When we arrived at the Indian camp, from its appearance and activity, the date could have easily been a century earlier. The hunt had been successful. Two moose had been quartered, a hide was stretched out on a spruce wood frame and two women were using flesh scrapers made from the femurs of one of the dead moose to flesh the hide deftly of fat. To maintain the angle of the scraper on the hide and to support their wrists, their hands were placed through loops of babiche connected to the moose bone scrapers. Nearby, two cast iron skillets were tipped almost perpendicular to the ground in front of a hanging fire. A giant patty of dough, defying gravity, was baking in each skillet. The dough was a kneaded mixture of flour, bear lard, baking powder, and salt. It was the makings of bannock, the staple carbohydrate of the North. On the lakeshore, an old Chestnut wood canvas freighter canoe was pulled up only halfway out of the water. The canoe's ribs were stained with moose blood. An old outboard motor, its cover gone was bolted to the canoe's stern. The motor was cocked up, prop exposed. Gas-oil mix was slowly dripping into a pool of water in the bilge and creating the colors of a greasy rainbow on its surface. We loaded the meat, some bedrolls and duffels into the plane. Before a half dozen of the men could hop in, Merlyn lingered for a slice of moose heart, some shredded dry meat and a hunk of bannock. The smell of woodsmoke, butchered meat, and two-cycle oil, permeated the air. It was a working camp, alive with humans carving a sustainable life from the bush and it was a good smell. I was quite surprised when, with all the talk, Merlyn made absolutely no mention of payment. Then after the flight and back at the float base, all Merlyn said to the guys was, "Put one of those hind quarters in the back of my truck." They did.

On the drive back to the Carter house on Vale Island we stopped at the West Channel and spoke with a guy I had never met. Merlyn said a few words and the old Cree hauled the moose hind quarter out of the truck bed and told us to come back in a couple days and he would have it all cut-up. No mention was ever made of the air charter bill or Jean's concern, Merlyn just said, "Boy, that's real good Rob. Nice moose meat for the winter."

George Marlow who passed away in 2016 lived in Snowdrift and was Chipewyan Dene. He was another best friend of Merlyn. Marlow was born in the bush halfway between Fort Reliance and the Snowdrift River. Merlyn met him in 1954. Merlyn was hauling fish from a camp on the eastern arm of Great Slave Lake, while Marlow was hunting caribou. It was back before the days commercial fishing was prohibited in the east arm. Marlow and his wife Celine spoke fluent Chipewyan and had their feet in both worlds. One was a world of independence that revolved in circles defined by the seasonal migration of caribou, the runs of fish, and the weeks when particular berries ripened. The other foot was in the recently unimaginable creature comforts of airplanes, snowmobiles, pickup trucks and flush toilets. Merlyn and Marlow could walk in both worlds.

The Chipewyan are some of the toughest, most independent people in the world. Since glacial ice retreated from the thousands of square miles between the Slave River and Hudson Bay, the Chipewyan and their ancestors had plied the territory by foot, moving with the migrating caribou herds and living a life of plenty on the tundra in the summer and the transitional forest also known as the "land of little sticks" in the winter. Nonacho Lake was a favorite wintering area of the caribou. Rings of rock once used to hold down skin tipis, stone knives, copper points and other evidence of First Nation camps can still be found on the landscape. The Chipewyan were among the last aboriginal Indian groups in North America to forsake their nomadic lifestyle and settle into villages across the northern rim of the provinces from Churchill, Manitoba to Fort Resolution, Northwest Territories. As with many aboriginal cultural changes, this happened more as a result of Government bribery and force than native inclination. In the 1950s there was a cyclical downturn in the population of the barren land caribou. A popular magazine story of the time highlighted a group of starving nomadic Caribou Inuit living in a desperate state near Ennadai Lake. Similar to the Chipewyan the Caribou Inuit's culture and sustenance revolved around the caribou. Possibly for the first time in their lives these people had been unable to find caribou, their staple. The families were starving. The immediate response by the Canadian Government was to move these people to areas of the Territories where caribou were still common. Merlyn helped with that airlift and tears came to his eyes when he remembered how emaciated and desperate his passengers were and how gratified he was to help them out. The magazine article put the eyes of the world on Canada. The negative view shamed mainstream Canadians and the government. How could a developed country allow a segment of its population to starve?

The government was motivated to do something, and in the ethnocentric norm of the times, houses were built in villages, and creative financial incentives encouraged the people to abandon their nomadic lifestyle and settle down. This applied not only to the Caribou Inuit but also the Chipewyan living only slightly further to the southwest. The Chipewyan, who remain fiercely independent and linked to the land even to this day, reluctantly adopted to town life, but still continued to live for long periods of time in the bush although less and less so as years passed. Much of Chipewyan culture still revolved around the caribou. As caribou are migratory, and settlements stationary, a problem arose. The people could no longer move by foot with the caribou who at seasons were now hundreds of miles from their villages. Chipewyan from Snowdrift could still reach the wintering caribou by dogteam and later by snow machines, but it was a logistical challenge unknown in the years that they had lived among the caribou. The Government, to ease this dissonance, began sponsoring "Caribou Hunts". Hunters were flown to frozen lakes where caribou were wintering and where the Chipewyan could set up camps. Caribou were shot and butchered, and the meat was flown back to the villages in a series of flights. Carter Air Service was often chartered to do the flying. The Chipewyan were not always quick to warm up to white people but affable, adaptable Merlyn won them over with ease. He was generous with his time, being quite happy to relax eating some caribou back straps and munching on a chunk of bannock while waiting for hunters to load the plane. It was also not lost on the Chipewyan that Merlyn could fit more caribou quarters in an airplane than any other pilot. Likely, it was even more important that Merlyn had already forged a relationship with Marlow and that relationship had included family gatherings. Kinship bonds are extremely important to the Chipewyan. In pre-contact times, marriages were even arranged to expand kinship connections. Peace keeping among the Chipewyan was similar to medieval Europe. When wandering bands of Chipewyan encountered other wandering groups, the key to keeping peace hinged on the elders from both groups finding some type of kinship relationship. If a connection could not be found violence and plunder could result. Although the Carters were not blood related, their social connection with Marlow and his family was highly valued by the band.

In the bush, the Chipewyan were comfortable and ran an organized hunt. Caribou were nearby and harvested by well-placed shots. The hunters usually targeted the cows as the meat was better and by early spring, calf fetus was an especially esteemed fare. Merlyn enjoyed the caribou hunting camps and his amiable acceptance was not lost on the Chipewyan who often felt judgment and prejudice from white people.

Merlyn's association with Marlow's family went beyond Government sponsored hunts. Merlyn liked a few caribou in the freezer every winter too. Although comfortable with a rifle Merlyn preferred the feel of a guitar in his hands. Merlyn never cared too much for hunting. He enjoyed duck hunting in Saskatchewan and some autumns he would drive with a few friends and a case of beer on the old highway to Fort Providence and shoot chickens (grouse) along the road, but he left the moose hunting and caribou hunting to

others. Dallas, Jean's younger brother, and Merlyn's ace p
most of the family hunting until a fatal car accident in 1973.
Merlyn and Marlow did go off caribou hunting together one late
summer day. Marlow shot a few caribou. Flying back toward
Snowdrift they encountered weather. Heavy rain and sleet reduced
visibility. High winds would have made a water landing at Snowdrift
dicey. Merlyn decided to land near the abandoned weather station at
Fort Reliance and camp for the night. The storm intensified and the
temperature was plummeting. Lacking sleeping bags, Marlow killed
two more big bulls and skinned them out in a unique way. With each
caribou, he made an incision along the inside of one of the legs, up
around the anus and down the other leg. Then with just a little
cutting, he pulled the skins off the carcasses as if he was taking off
giant socks. By keeping the hair turned inward he had created two
cozy sleeping sacks. The two friends slid into the skin sacks and
slept side by side like toast in a toaster. They awoke refreshed in the
morning having been warm and dry all night. But Marlow warned
Merlyn not to try this with a moose. Moose hide is so thick that if it
froze overnight it would be impossible to cut and would encase the
occupant in a hairy tomb. With a moose it was better to cut the hide
into three blankets. Merlyn was as comfortable sleeping inside a
fresh killed caribou skin as he was in his Maui timeshare condo.
More than once I heard him say while standing in front of his wood
stove in his Vale Island home, "We've got a warm fire and grub to
eat, what else is there, Rob?"

One day, Merlyn had dropped Marlow off with a couple
Chipewyan friends to hunt muskoxen. He promised to return to pick
him and the animals up later. He told Marlow that if the hunt was
successful to prepare some muskox ribs for supper. Merlyn was
delayed hauling fish and Marlow's friends were anxious as darkness
loomed. They were unprepared to spend the night and began to
assume it was too dark for Merlyn to land. But Marlow had
confidence and insisted they maintain a big fire, not to stay warm,
but as a beacon for Merlyn. Sure enough, the throbbing rotary engine
of Merlyn's Single Otter pierced the inky black sky. In came Merlyn,
and as soon as the plane had landed smoothly on the frozen lake,

Merlyn went straight to the campfire and tore off a musk-ox rib, gnawing at it with gusto. Although aboriginal hunters, by treaty, were granted almost unlimited hunting privileges, resident white people could also buy hunting licenses and within liberal seasons and bag limits harvest a broad range of wild game. In the 1970s the limit was four caribou per person. Barren ground caribou is a delicious, low-fat meat which compares favorably with the best grass-fed beef. The huge Beverly herd of caribou, estimated at over 425,000 animals in the 1970s, spent their summers far to the east on the barren lands, but wintered in an area centered around Nonacho Lake only 215 miles from Hay River. There are no all-weather, or ice roads, within almost 200 miles of Nonacho Lake. The only practical way for hunters to get there is by bush plane.

This kind of meat procurement has more in common with going to the supermarket than sport hunting. It is not intended to be a "sport." People who do this work, do so to put meat on the table. There is no interest in making this a "fair chase" experience. As one meat hunter once told me, "The best view of a caribou is through the sight of a gun." With that mindset, the easier to attain that vision the better. Caribou have evolved for thousands of years to avoid their arch enemy, the arctic wolf. In the bush with trees and deep snow, it is easier for the wolves to ambush the caribou than out on an open lake where with a head start even a young caribou can outrun a wolf. When threatened caribou run from the bush out on to the frozen lakes. When a plane flies over the transitional forest it alarms the grazing caribou and their response to alarm is to run to what always has been the safety of the frozen lakes. For airborne hunters, this is a perfect situation. As the plane flies over the lake the caribou run out from the bush and stand on the lake in scattered herds. Caribou are spotted from the air and the pilot lands the plane on the lake as close to the herds as possible, sometimes almost colliding with the animals on landing. The shooters jump out and accurately shoot a dozen or more animals. The plane is then taxied from carcass to carcass. Sport hunters usually gut the animals and then throw gutted carcasses in a pile in the back of the airplane cabin and fly back to Hay River. This is how it usually works.

But Merlyn told me an amusing story. Weather can be quite different in Hay River than 215 miles to the East. When Merlyn was asked about a typical Nonacho Lake winter he would always say, "50 below every February day." I lived at Nonacho Lake one winter and he was not far wrong. One warm February day in Hay River three hunters bugged Merlyn about taking them caribou hunting at Nonacho. Dressed for a mild February day and riding in the heated Twin Otter (the Single Otter heaters barely kept the plane above freezing, and on some flights Merlyn would light a plumbers blow pot inside the plane for heat!) the crew was unprepared for windblown snow and fifty below at Nonacho Lake. Other than the weather, everything went as planned. As the plane circled above the lake, caribou paraded out on to its surface in droves. Merlyn landed the plane and hunters jumped out of the door like Marines. Three rifles pierced the frigid air as ears rung. In a moment it was over, and more than a dozen dead caribou were scattered around the lake. After the adrenaline of the shooting had begun to wear off, the cold penetrated the thin parkas of the hunters and seeped into their bones. Merlyn suggested that the hunters forgo the gutting until they were in the heated hangar back in Hay River. He would just taxi the plane from carcass to carcass while the hunters threw the animals whole into the cabin. They would be back to Hay River in ninety minutes, long before the meat could spoil. So that is what they did.

2,000 feet above ground heading west toward Hay River at 175 mph all was well. The hunters sat and stared at the pile of steaming dead caribou, blue tongues hung listlessly, glassy eyes vacant. Then one hunter elbowed another and in a voice that could be heard over the whine of the turbines said, "I think I saw an eye twitch on that caribou over there, that big bull." A minute later there was no doubt. The caribou bull first lifted then shook his head, large antlers rattled against the walls of the cabin. Apparently, the shooter's bullet had just grazed the skull of this animal rendering it temporarily unconscious, but far from dead. The beast was confused, but struggling to stand up. Before one of the hunters could load his rifle, ɔecame aware of the commotion in the cabin and turned, ᵬ⸗ ᴉt each hunter and at the increasingly agitated caribou. "Do not fire a rifle in my plane!" So, the three guys had to wrestle the beast, an animal that can cave-in the ribs of a 120-pound arctic wolf with one kick, to the floor and slit its throat with a knife. When they arrived in Hay River, the Twin Otter cabin looked like a crime scene. The walls, the windows, the hunters were all splattered with blood. It was just another day of work for Merlyn Carter.

This same event could not be repeated today. Most caribou herds around the world are in steep decline. The Beverly caribou herd that once numbered over 425,000 animals is now extinct. There are many possible explanations for this catastrophe.

Caribou populations have always been cyclic. Dips in their numbers have caused hardship and famine for the people who depend on them — as they did for the Caribou Inuit in the 1950s. What has made this down-cycle such a catastrophic one? It appears that a natural dip in population has combined with other factors that may spell the coup de grâce of the barren-ground caribou based ecosystem.

The most plausible contributing factor appears to be a warming climate. Long-term Northwest Territories residents have repeatedly told me that they now experience warmer winters, especially warmer low temperatures at night, shorter frigid cold snaps, thinner lake ice, earlier springs, melting permafrost, later freeze-ups. Winter ice roads have a shorter season now. Animals such as whitetail deer are being seen farther north. Birds such as magpies that even lack Chipewyan names are being sighted in places they've never been seen before. Scientific data supports this anecdotal record. How does this affect the caribou populations?

One theory is that the earlier and earlier spring break-ups are resulting in a drier and drier transitional forest. The transitional forest is essential for wintering caribou. As the forest has become drier, it is more susceptible to wildfire. The successional vegetation that follows a wildfire, although nutritious for moose, is not satisfactory for caribou. The lichens and sedges required by caribou take many years to regenerate. In recent years, wildfire has consumed much of the wintering range of the barren-ground caribou. The fact that the most southerly caribou herds are in the steepest decline supports this theory. The Bathurst caribou herd just north of the Beverly herd has also incurred a precipitous decline going from estimates of 350,000 animals in the 1990s to under 20,000 in 2015.

Caribou cows that are undernourished during the winter months will re-absorb their calves. This has been documented by a lack of caribou fecundity on the calving grounds. Without many new calves, attrition from wolf predation and human hunting, causes an ever-increasing population decline. This phenomenon is exacerbated by the fact that barren ground caribou are herding animals. The health of the individuals depends on the health of the herd. The mass drop of calves on the calving grounds is an adaptive strategy. Caribou herds, especially the pregnant cows, move as a large unit to a calving ground. Wolves that follow the herd straggle behind because they need to stop to den and whelp their pups. In past times, by the time hundreds of thousands of caribou reached the calving grounds there were relatively few wolves. As the size of the herd declined the ratio of wolves to caribou increased, further impacting calf survival.

There are other factors too. Vegetation adapts quicker to a changing climate than animals do. Scientists surmise that because the tundra is warming earlier, plants that were once in their nutritious sprouting stage when caribou arrived, are now in a later and less nutritious flowering stage. This change in nutrition too may affect the ability of the cows to birth successfully and raise calves.

It is also possible that a brain worm which infects whitetail deer is now being spread to caribou as their ranges have merged. The deer have adapted to live with this parasite. It is however, fatal to caribou.

Some scientists have suggested that an increased black fly and mosquito population has also undermined the health and reproduction rate of the caribou. This theory seems less plausible, as anecdotal evidence of pilots and canoeists does not suggest any major increase in insect levels.

Although much has been written about the threat the changing climate poses to polar bears, somehow barren ground caribou have missed the spotlight of media attention. As tragic as the

rance of polar bears would be, their demise pales in
ison to the loss of caribou. Polar bears are an apex predator.
Caribou are a primary link in a food chain that impacts many other
species. Wolves depend on caribou. As caribou disappear so do the
wolves. Arctic and red fox, wolverines, and possibly other predators
scavenge on wolf killed caribou. For bald eagles to have enough time
to incubate their eggs and for their offspring to fledge and be ready
to fly south, they must nest early. Primarily fish eaters, bald eagles
are dependent on wolf killed caribou to survive in March and April
when the lakes and rivers are locked in ice and fish are inaccessible.
Gray jays and ravens also follow the caribou herds, feeding on kills.
Barren ground grizzlies depend on young caribou for their spring
diet. In the early years of the Beverly caribou herd decline, grizzly
bears were observed near Fort Smith, far out of their traditional
range and possibly searching for something to replace their diet of
caribou calves. These bears have since vanished. Traveling the
barren lands in summer or the transitional forest in winter, these
days, where once caribou was king, I am more likely to see a muskox
or a moose than a caribou. Wolf dens are empty and the landscape is
remarkably quiet. Caribou antler sheds that in the 1970s littered the
eskers around Nonacho Lake like seashells on a beach are now
absent. It is flabbergasting to scientists, aboriginals, and sport
hunting outfitters that these huge caribou herds could disappear so
quickly.

One more contributing factor to the Beverly caribou herd's
demise was that modern technology kept harvest numbers high even
as the overall size of the herd drastically declined. The wintering
density of caribou herds remain constant even as their overall
population decreases. The location of the Beverly herd in the 1970s
would be like a dinner plate on a map, but as years passed it would
more closely resemble a saucer or a poker chip. The density
remained constant, but the area the herd occupied at any particular
moment continued to shrink. Even today, with caribou at record low
numbers, hunters can still find a lake in the Bathurst caribou herd
wintering grounds where animals appear abundant. Northerners, both
aboriginal and white, distrust scientific reports. Dire fears in the past

of disappearing caribou herds have proved specious. For many years, it was difficult to convince aboriginal or white hunters, who could still find places where they were surrounded by caribou, that the decline was as serious as it is.

In the past, where caribou suffered a cyclical downturn in their numbers, hunters could not find the scarce pockets where the animals still thrived. Just a few years ago hunters could track GPS equipped caribou collars on a government Internet site to ascertain the exact location of the entire diminished herd. So instead of the caribou being able to hunker down and recover, the hunting pressure actually increased.

Ironically the recovery of the Bathurst herd may be determined by aboriginals. Ironic because aboriginal hunting did not cause the decline, but the First Nation people may need to restrict or even eliminate caribou hunting altogether if the Bathurst herd has any chance to recover. This is a huge economic and cultural blow to the people. For countless generations, their culture has revolved around many aspects of the barren ground caribou. The Chipewyan were even once called the "caribou-eaters." Barren ground caribou were also integral to the Dogrib and Yellowknife Dene cultures. It is impossible to comprehend how devastating the demise of the caribou will be to their culture and historical traditions.

Politically, there are also challenges that has slowed the response to the crisis. The Beverly caribou herd spent time in Manitoba, Saskatchwan, Alberta, the Northwest Territories and Nunavut. All these government's natural resources departments were discrete from each other and under different political jurisdictions. Native communities that hunted the herd also came from many different bands. A coordinated collaborative effort was ground breaking, time consuming and remains to this day daunting.

The Beverly herd is gone, but efforts are underway to save the Bathhurst and other herds across the North American arctic and sub-arctic. Increased cooperation between First Nation bands and more accurate and focused scientific research are hopeful signs, but the future of the caribou is uncertain.

Barren ground caribou waiting to cross the Mara River
(photo by author)

Chapter 11
Ice

As an only child, with a disarming smile, and a "who me?" kind of feigned innocence, Merlyn usually got his way. That did not change as he grew older. He had an uncanny ability to seemingly go with the flow, but at the same time get what he wanted. It wasn't that he avoided conflict; conflicts just seemed to avoid him. He was more a lover than a fighter and his childhood chums, as well as his adult friends, employees, business associates, children, and grandchildren, could not recollect a single instance of having a quarrel with Merlyn or even hearing him speak in anger. When backed into a dilemma, Merlyn always responded with kindness. Accuracy was secondary.

Merlyn gained a lot of weight in his middle age, and Jean was frequently putting him on a diet. She carefully poached trout instead of frying, and limited his supper portions. Merlyn obediently went along with the restrictions, at least while Jean was watching. Many evenings while on his strict diet Merlyn would take me along in his pick-up truck when hauling a load of garbage to the dump, only to stop at the Pizza Patio on the way and pick up a pizza. Merlyn would share the pizza with me, although I knew he was capable of devouring an entire, large, after supper, pizza on his own. After consuming the pizza, and always with a laugh, he would say, "Make sure we throw that pizza box out at the dump with our junk. We don't want Jeanie in on this one."

I wasn't alone as a snack conspirator on runs to the dump. During a summer when Merlyn hired Gordon Clark to fly his Single Otter, Clark can remember clandestine dump runs. They would stop on the way at Godwins' supermarket to pick up a half-gallon of ice cream and two spoons.

Myles remembers a time when he was drinking with his dad out in the den of the family home on Vale Island. Merlyn was pouring drinks from a bottle of rye. About half past midnight Jean came out of the bedroom and sharply said, "Enough of that drinking." She grabbed the bottle and poured what was left down the kitchen sink drain. Another man might have been angry; another man might have gone to bed; all Merlyn did was to whisper to his youngest son, "Just be quiet for a few minutes Myles. I have this figured out. When I went to the liquor store this afternoon, I bought TWO bottles, I've got the other one hidden in the woodpile. As soon as Jean gets back to sleep, I'll pull it out." That's what he did. Twenty minutes later Myles watched him retrieve a second identical bottle from the woodpile and say with that endearing smile, "Now, let's get back to that little party." It was hard to determine whether Merlyn got more pleasure out of the drinking or the duplicity.

As a storyteller, Merlyn had a sense of timing that could make a mundane joke seem fresh. He had the gift that even if I had heard the story a half-dozen times before, it seemed just as funny the seventh time. Somehow, he did it without swearing. Language on the northern frontier was, "colorful." Merlyn was surrounded by men who dropped a swear-word between each conjunction. Yet Merlyn rarely swore. Sometimes his stories had a ribald theme, more likely not. When confronted by a surprise or challenge he would more often respond with "Holy Smokes!" than an F-bomb. Merlyn was rough on equipment, and when it broke, he would as likely laugh, make a face, and say, "I guess that's had the biscuit." He never seemed to second guess himself.

Hot air is less dense than cold air. It is a big challenge for a pilot to get a heavy load in the air on a sultry day. This is as true with modern airliners as it is with bush planes. As commercial airline routes expanded more and more to Sunbelt and hot Mid-eastern destinations and passengers got fatter and fatter, jumbo jets have needed to be fitted with more powerful jet engines. For float pilots flying overloads, summer days meant getting an early start to avoid the heat of the day or waiting for an afternoon breeze to create a bit of lift and more importantly put a little chop on the water to help bounce the plane off the water while it was planing "on the step."

Merlyn loved the challenge of getting an overloaded airplane airborne. His engineer for several years was Dave Cathcart. It was a hot summer day, with not a breath of air and the water was like glass. Merlyn had an overload on the Single Otter, and after two failed attempts to get off, he refused to unload any of the cargo but instead gave it one more try. After a two-mile run on the river, the plane would not break off the step. Finally, the prop seal failed and blackened the windscreen with oil. Merlyn taxied to the dock shaking his head. When Cathcart pulled out the damaged prop seal, he had never before seen a seal so deformed by heat. Merlyn remarked as he peered over Cathcart's shoulder, "Looks like that one's had the biscuit."

Once while out at Nonacho in the midst of a cabin construction project, the fuse on his diesel generator kept blowing. He finally replaced the fuse with a steel washer. After 30 minutes, the generator was fried. Again, with a shake of his head, he went out to the shed to find rusty hand tools to use to finish the job. He dragged that busted diesel powerplant out into the bush with his old farm tractor. It sits there today. Back in town, he ordered a new one with three times the amperage capacity of the old one. Hauling that behemoth to Nonacho, sticking half out the cargo door of the Single Otter, was a miraculous feat. Already overweight, Merlyn had to put more even more load on the opposite side just to balance the weight so he could keep the wings level.

Merlyn could push equipment to their limit and beyond, usually without consequence. He would run his Ski-doos up the sandy beach in summer to the shed to fetch another case of beer or load his Single Otter until the floats were barely above water. He knew how many rivets on the top of the float could be submerged to still get the plane up and off the river. Getting overloaded planes airborne and flying a heavy load takes extraordinary skill. The plane lumbers along plowing through the air.

The aviation term "stall" is sometimes misunderstood by non-pilots. Stalling in an airplane is totally different than stalling in a car. In a plane, it does not mean the engine quits. When a plane stalls, it is because the wings "quit." Usually, this is because the angle of attack of the wings has become so steep that smooth airflow, and lift, is disrupted and the wing drops, sometimes quite sharply. Airplanes have a minimum air speed, below which they can no longer sustain flight. This is called the stall speed. A stalling airplane is a falling airplane. Different wing designs create different outcomes in a stall, but all airplanes lose altitude in a stall, some faster and wilder than others. To recover from a stall, the stick must be pushed forward. Unless power is applied simultaneously, a stall will result in a dramatic loss of altitude.

When flying an overload, the stall speed of the wing and the airspeed at cruise narrow. It takes a light, careful touch to avoid calamity. Merlyn had great flying sense and never unintentionally stalled an airplane. Merlyn also knew how to load an airplane as well as anyone, and usually got his planes up off the water regardless of load. He never lost a passenger, nor was ever a pilot or passenger hurt when Merlyn was pilot-in-command. His three serious crashes, flipping the 170B, flipping the Lockheed and going through the ice with the Single Otter all happened with underloaded, not overloaded aircraft.

There were residents of the North who would fly with none other than Merlyn at the controls, no matter how many people were crammed on the plane.

In his pockets, Merlyn always carried a pair of pliers, an old screwdriver, a roll of tape, a little book, and a wad of cash. To get out of a jam, he seldom needed anything else.

By the 1980s, winter fishing had slowed down dramatically and fish flying was correspondingly slow. Merlyn had a clause in the contract with his insurance company that if his planes sat idle and were not used for a period of weeks, he would get a rebate on his policy. That didn't mean Merlyn did not fly.

If he did some flying for a Chipewyan caribou hunt, a few fish hauls, or some government flying, he would only keep track of his flights in his little breast pocket notebook, not in his log book. "If I have a problem, well then I will fill out the log book. But if I don't, well then, no sense doing all that paperwork or paying all those premiums." This procedure makes it hard to determine just how many hours Merlyn flew on skis.

Merlyn was not alone neglecting his log book. Perry Linton called log books the worst enemy of a flying service. "Just when you would get an engine broken in and running good you would max out the hours and be required to put in a new engine that often was junk."

It was impossible to build a successful northern flying service by always following the rules. The Twin Otters were the first turbo prop planes Merlyn owned. Turbines ran just as well on domestic heating oil as jet fuel. More than once Merlyn would have the local heating oil delivery truck fill his bulk fuel tanks at the float base. That is not to say he did not care for his airplanes, but if fuel oil burned as well in the turbos as jet fuel and cost half the price, why not? He often repeated, "Take care of your airplanes, and they will take care of you." Merlyn just wasn't stupid about it.

That attitude extended to his long-term engineer Norm Weber. Weber had a bit of a drinking problem. I cannot say that I ever saw him sober, but he was a solid engineer and an absolute master at fabric work. Merlyn usually steered clear of fabric airplanes because even with a guy like Weber on staff, fabric airplanes require more maintenance and care than all-metal planes. That said his two Norseman aircraft were all fabric and the control surfaces on his DC-3 were fabric. I never asked Weber to do any work on the fabric of my Piper Pacer XXO because Merlyn kept him busy and I didn't want to contribute to Weber moonlighting on my account. However, when my annual Certificate of Airworthiness (C of A) was due, I asked Merlyn if I could get Weber to do the inspection for me. Merlyn said, "Yeah, Rob just bring him a bottle of rye and tell him I said it was okay."

I did as Merlyn suggested. I remember Weber just licked his lips when he saw the bottle. He opened my log book and with a shaky hand signed off the inspection. I waited a moment and said, "Is that it?"

Weber said, with steely eyes, "Yeah, kid that's it."

It is safe to say Merlyn did as much northern bush flying as anyone. Partially, this was due to his longevity. Once while we were sitting around the dining room table of his home, he recounted how many of the pilots that he had started out with in the 1950s had died, many in fatal crashes. Also, many of the pilots Merlyn had hired, had taken jobs in the south, doing "blue suit" flying, and left the blow-pots, float pumping and loads of dogteams to Merlyn.

Joe McBryan was an exception. After getting his start with Merlyn in the early sixties he went on to build Buffalo Airways into the most unique flying service in the world. McBryan operates a fleet of mostly 1930's DC-3s and supplies mines and exploration companies until they are able to build airstrips that modern freight hauling aircraft can handle. He may be running the only scheduled DC-3 service left in the world — a daily run back and forth to Yellowknife from Hay River. McBryan says, "Merlyn absolutely flew more jumbo whitefish than anyone who has ever lived."

When I first put floats on my plane I needed to get a float rating. I could accomplish this by logging in some air time with a commercial float plane pilot and having him sign me off. Landing, taking-off, and taxing on water requires a different skill set than land flying. I asked Merlyn if he would help me out and, of course, he said, "Yes."

It was a brisk spring day when Merlyn joined me in XXO for my float endorsement. I flew one circuit off the river with Merlyn sitting in the right seat. Merlyn then says to me, "Rob, you're scaring me a little bit, but you know, you got a small airplane, and it will fly way better without me in it, just let me out at the dock and I can watch you from the ground." I did what he asked and dropped him off. On the very next circuit, I looked down and noticed that he had already gone into his office. Later that day he signed my log book. It may have been the quickest float endorsement in the history of Canada.

When I spent a winter at my cabin on Nonacho Lake, my nearest neighbors to the north were 90 miles away in the small Chipewyan community of Snowdrift (now called Łutselk'e). To the south was the now ghost town of Uranium City some 110 miles away. To the west was Fort Resolution 150 miles away. To the east, basically forever, maybe Eskimo Point (now called Arviat) on the Hudson Bay. Even so, I remember once at Nonacho Lake when the camp ran out of beer, Merlyn said, "Uranium City is only 110 miles away, I could be back in two hours." He was serious, but he didn't go. He had already drunk a few beers, and although Merlyn bent the rules a bit, he was a careful pilot. When I asked him once, if in his younger years, just for fun, did he ever fly under a bridge? He said, "No, I didn't do stuff like that and you know, a lot of my pilot friends that did, are dead." Jean chimed in from the kitchen, "There are bold pilots, and there are old pilots, but there are no old, bold pilots." But then Merlyn leaned over to me and said when that when he was in his early teens, a buddy's father had a J3 Cub. His buddy knew a girl on a ranch not far from Meadow Lake that was pretty wild, they borrowed the friend's dad's plane and flew over for a long visit. They landed the J3 in a rough horse pasture adjacent the farm house. It was a pleasant afternoon but made less so when they returned to the plane to discover that a pony had eaten much of the fabric off the wings. Fabric airplanes can be fussy. Ravens can be as bad as ponies. Something about the dope used to coat the fabric attracts a variety of animals' attention.

In the winter, I kept XXO parked at the Fort Resolution airport. Ravens, until I shot one and left it hanging on a wing strut, would sit on the wings and pick at the fabric causing a hundred dollars of damage with each tear. It was tempting to bring another bottle of rye to Weber instead of hiring the repair out, but I never did. Merlyn and Jean had helped me get my start in the Northwest Territories and you never forget or betray those that help you when all seems lost. If a pilot was going to be a successful northern bush pilot in the last half of the twentieth century he had to take some risks, bend some rules and take care of his airplanes. If you want to be successful in life you have to take some risks and be faithful to friends. A little luck also helps.

As for flying into Uranium City after drinking a few beers? Merlyn could hold his liquor better than any man I knew, but he was careful about flying and drinking. He didn't do it very often.

"We are missing a nut," Cathcart said with a sigh. It was the fourth day the Lockheed had been grounded on Lac Grandin just southeast of Great Bear Lake. The temperature at night was 55 below, in the short December day it had warmed to minus 48. "There must be a nut on this plane that's not holding much." said Merlyn, "I'll go look." He took a pair of pliers and a screwdriver. It was too cold to operate a Crescent wrench. Anyway, Merlyn could do just about anything with a pair of pliers and a screwdriver. Merlyn found a nut that wasn't holding anything too important and they were back in business.

Merlyn had landed almost a week ago with a flat tire on the Lockheed. Unperturbed he had loaded 3500 pounds of fish inside the cabin knowing with one good wheel he could still get airborne. But after warming himself up in a tent with the crew of fishermen, he turned to the Lockheed only to see the other tire had flattened. He would need help to bring out a couple of new inner tubes. All afternoon he couldn't raise Yellowknife on the HF radio, so he ate dinner in the cook tent with the boys, pulled his bedroll, a Woods Arctic Three Star Eider Down, from the plane and made plans to bunk in the fishermen's tent. This was one of those nights when Merlyn didn't come home, and Jean, back in Hay River, could only wonder and worry. He had left that morning sick with the flu and the coldest snap of the winter gripped the North. This is the life of a wife of a bush pilot. Never sure what each day will bring and whether your guy is safe or in strife. A pilot's wife is the life Jean chose and did not question. But the ring of a phone even ten years after Merlyn's passing still makes her short of breath. But on this evening, Jean hoped the phone would ring as she busied herself frantically cleaning the house knowing she would not sleep until she heard why Merlyn was overdue.

Four-hour days and long nights are what the climate deals in December in Great Bear country and the nights seem interminable in a canvas wall tent. The tin wood stove is stoked tight with spruce and when it starts to roar, it kicks out the heat. Sometimes so much so that the stove pipe glows cherry red. At these times, the temperature in the tent soars and the boys crawl out of their bags in a sweat. As the embers start to burn down and the sheet metal stove starts to bang as it cools, everyone crawls back into their bags for an hour of sleep. That is until someone gets too cold to tolerate the frigid air and stokes the fire and the cycle repeats. At 55 below zero, the night is a half-wake, sequence of hot and cold.

In the heat of one cycle, Merlyn notices the Aurora has gone dark, and he decides it's worth it to trudge the quarter mile across the squeaky snow to the plane and try the HF radio. Without the electrical interference which the Northern Lights can sometimes create, maybe Merlyn can get a transmission out. He trudges across the snow to the plane, carrying the warm battery as his nose hairs crackle and freeze. By the time he hooks it to the HF radio his fingers are numb stubs. He plunges them back into his mitts. The padded seat is frozen and feels as if it is cold concrete. The radio signal pierces the clear air and he reaches Yellowknife with a message to be relayed by phone to Jean.

At first light, a Cessna 180 on skis will leave Hay River with Cathcart and the tubes.

It will be afternoon before they arrive. Back on the lake his load of fish is frozen, worthless. The crew cast the fish out on to the ice, a dozen ravens pick at them. It's already getting dark when the whining Cessna pierces the blue steel sky. Too cold to fish, one man with a hand black from frostbite returns to Hay River in the 180 and the rest haul firewood from dead spruce trees along the lakeshore or wait in the tent with a rifle, hoping a wolf will try to sneak a dead fish from the heap before darkness deepens. Why shoot a wolf? Parka trim is worth a couple bucks a foot. The cold snap persists for days and with just a few daylight hours to work, the repair is slow. Landing on the ice with a frozen deflated tire caused some damage to the wheel, and that too must be repaired. The big spongy tires are one asset that helps make the Lockheed such a great winter bush plane. Only a bit larger than a Beech 18 it can carry almost twice the load of fish. Part of the reason was the high and broad undercarriage and the 35-inch in diameter tundra tires. These tires had only a six-inch rim which made them difficult to change. The wheel had to be removed and that meant each side had to be jacked up off the ice.

Even the simplest task becomes exponentially more difficult in frigid conditions. They had to haul the wheels up to the hot tent, thaw them and install new tubes. When the cold snap finally broke, nets could be reset; engine tents were draped over the engines and blow pots fired up to heat the frozen radial engines. Soon there was another 3500 pounds of fresh fish on board and fully inflated tires below. Merlyn had an easy manner with fishermen and enjoyed sharing a meal together, but he seldom slept over at a fish camp. A side of Merlyn was like his father, he never completely embraced bush life. He loved to eat caribou but preferred to eat it at home and after a big dinner crawl into a warm bed with Jean. He would rather do the flying, and delegate someone else to do the blood work and any aircraft repairs out on the ice. Had he known how long it would take to fix two flat tires he would have gone back to Hay River in the 180 and returned when Cathcart had finished the job. Sometimes you don't know, and stuff happens. You just have to barrel on through and make the best of it.

As often as planes went down with mechanical or weather issues, it seems like a lot of attention would be paid to the survival kit. In Merlyn's case, this was not necessarily true. Weaned in an era of "underpowered" airplanes hauling overloads of fish, every extra ounce meant lost revenue. Any unnecessary weight was stowed back at the base. Even painting an aluminum plane was second-guessed, because of the weight of the paint. Merlyn became accustomed to flying, literally by the seat of his pants and with barely a pair of pants in the plane. Jean did believe a survival kit was a good idea and she would painstakingly assemble kits which consisted mostly of food.

Although many flights were scheduled well in advance, a bush pilot is at beck and call. It was more the rule than the exception when eating dinner at the Carter's for the phone to ring and for plans to immediately change. I can remember several instances just as we were sitting down at the table for a feast of trout or caribou like only Jean can cook, to have the phone ring. If it was flying business, Merlyn would excuse himself for a flight of a few hours or maybe a few days. Abrupt changes in plans were so accepted by both Merlyn and Jean that they were the norm, not the exceptions. Maybe "excuse himself" was an overstatement, he would just get up and go. No second-guessing, no apologies, just get ready and go. Forest fires, lost hunters, fuel hauls, can't wait for tomorrow. The gist of this was, and much to Jean's chagrin, Merlyn sometimes left for flying with an empty stomach. Without hesitation, he would eat food out of the survival kits for a snack.

In 1997, I was on a canoe expedition with my niece Karen. It was a hard trip over 400 miles up the Elk River and down the Taltson River to Nonacho Lake. We were three days overdue. Merlyn and his son Dean were becoming concerned. They flew our proposed route backward from Nonacho in TWN, Merlyn's last Cessna 180. They found us at the top of Gray Lake. Being overdue, our provisions had run low. Our daily rations consisted of boiled fish and oatmeal. Merlyn noticed we were a little lean and asked Dean to pull something out of the 180. Dean returned with a couple of small cans of Dinty-Moore Beef Stew which was all that was left in the survival kit. We removed the lids and ate the contents cold, with our fingers. Merlyn watched but didn't say anything, even when I used my fingernail to capture the last bit of gravy from the groove of the can. He did offer to tie our canoe to a float and fly us the last 50 miles to Nonacho. Four people, camping gear, and a 17-foot canoe on the float would have been an overload. I had no doubt Merlyn could do it, but we declined. We wanted to complete the trip by canoe.

Six hours later Karen and I were camped on a beach next to the delta of a small unnamed river. We heard the same familiar whine of the Cessna 180. His unexpected morning visit was the only plane we had seen in three weeks of travel, so we knew this was Merlyn coming back. Earlier that day, without saying a word, he must have noticed our hunger. He opened the door of the plane and brought out caribou stew, Jean's fresh baked bread, cookies, fruits, beer and two wild blueberry pies. We sat together and shared stories. Karen and I had a run in with a grizzly on the portage between Rennie Lake and a tributary of the Taltson. Readers familiar with Alaskan grizzlies may discount a grizzly encounter as ho-hum. But the bears of the Territories, both grizzlies and young black bears, lack the good nature of the fat salmon-eating Alaskan bears. The bears of the Canadian Arctic are accustomed to eating warm-blooded animals — any warm-blooded animal will do. The bear that charged us by Rennie Lake was hungry and had we not been carrying a shotgun or had we even been facing the opposite direction while eating our breakfast, I have no doubt this book would have never been written.

Merlyn told a story about how Eugene McKay had shot a big bear at Nonacho. It had charged him and his grandson. McKay had shot it five times with his 30-30, then emptied his grandson's .22 into it and finally killed the bear with his axe. When he skinned it out, the silver tipped furred hide covered the entire floor of Merlyn's shed at Nonacho. Eugene sold the skin to one of Merlyn's pilots for $50. The pilot traded it to the owner of the hardware store for a little black and white TV. The hardware store owner sold it to one of the NHL hockey players coming in from Nonacho for $500. Merlyn laughed, McKay didn't make out too well on that little deal and then laughed at himself when he said, "I purchased the Norseman including floats and skis for $11,000 then sold it a few years later for $20,000. You know, I thought I made a fortune. Now they sell for over $400,000."

He also mentioned that circling the river looking for our campsite that afternoon, he saw a waterfall he had never noticed before. It was just a few miles upstream from the creek we were having our little picnic on. And then he jumped back into his plane and headed back to Nonacho.

Over a hundred miles of flying with aviation gas (avgas) that had to be flown in and worth at least ten dollars a gallon, and Merlyn's time — all because a couple of his friends were running short on provisions. Merlyn and Jean did this stuff every day of their lives. When I nominated Merlyn for the citizen of the year back when I lived in Hay River in 1982, a member of the Town Council scoffed, "Merlyn Carter? He is not even in the Lions Club." But the truth is when someone was lost in the bush, needed a helping hand, a part-time job, a few bucks, or a discounted flight to a trapline, Merlyn and Jean found a way to help, and they did it quietly. And they did it regardless of whether the person was Slavey, Chipewyan, Cree, Metis, an old-time resident, or a visiting tourist. They helped out those in need and never second-guessed their generosity.

*Merlyn on Gray Lake, TWN in the background. Bringing two
hungry canoeists a feast of Jean's home cooking
(photo by author)*

It might be tempting to discount the Carter's generosity because they were so rich, that they were so flush with cash that they could afford to throw it around. It is a misconception that as owners of Carter Air Service, the Carters were super wealthy. No question they had some good years but they also had some years where they were nearly bankrupt from airplane maintenance costs, wreck repairs, exorbitant insurance premiums, mods to meet Air Transport regulations and just the fact that, after the booming commercial fishing years passed, for many winter months their planes sat idle. The Carters were generous with their time, money and friendship because they cared. As old time Hay River residents, they were not alone in this virtue. Caring, helping out, was a character trait of Hay River residents. At the Hudson Bay Store buying provisions for a canoe trip my niece, Karen, was short almost ten bucks, a voice from the back of the line from a stranger to Karen said, "I'll cover it."

The reason survival kits in Carter aircraft were piddly or non-existent was because of Merlyn's insatiable appetite, and his history of flying fish where every pound cost you a dime. Both mindsets never changed.

Perry Linton who went on to build his own successful charter and outfitting business in Norman Wells started out working with Merlyn. In fact, Linton's first flight in an airplane was when Merlyn gave him a ride from Fort Providence to Yellowknife in the Cessna 170B on floats. Linton's first solo commercial flight was in Carter's float equipped Cessna 180 KOW. Merlyn was taking some sport fishermen up to Point Lake and needed a second plane to carry provisions and gear. Merlyn told Linton just to follow him to Point Lake about 300 miles north of Hay River. When Linton got to the float base in Hay River, Merlyn had KOW all fueled up and filled with gear and provisions. Merlyn would take the passengers in his other Cessna 180 OLD. The take-offs went smoothly, and both planes were soon headed north. When a pilot is following someone in an airplane, they just get a few hundred yards behind and mindlessly follow the leader like a calf following a cow. After they both landed on Point Lake, unloaded and set up camp, Merlyn told Linton that he was going to check in at another one of his camps on

Contwoyto Lake and for Linton just to head back to Hay River in the empty plane. Global Positioning Navigation Systems (GPS) had not yet been invented. KOW did not have even an Automatic Direction Finder (ADF). The panel was equipped with just the most basic navigation aids: a compass, altimeter and a directional gyro. As Linton taxied out on to Point Lake, he looked in the pockets for a map. Nothing. He looked back behind him in the cabin for a survival kit. Nothing, not even an axe. But as Linton says, "What good would an axe be anyway?" the terrain around him was treeless tundra." Linton wasn't too flustered, he flew the empty plane south, and in an hour, he could see Great Slave Lake and from there it was easy (relatively easy) to find Hay River.

The best survival kit was in the mind of the pilot, and Merlyn knew how to survive. Whether it was opening a bottle of beer with a pair of pliers or fixing a tractor engine's rocker arm with a hose clamp, Merlyn could do it and he knew he could do it. Alone with TWN, in the current of the Taltson River, he had a dead starter. Despite the plane being on floats, despite the fact that the plane's owner's manual states that it is impossible to prop (start the engine by swinging the propellor by hand) a 180, Merlyn did exactly that, the impossible. Merlyn might just say, "What's an Owner's Manual?"

Merlyn did like to carry a gun in his planes. Pictures from the sixties show him wearing a holstered Smith and Wesson .38. He stashed a little AR-7 survival .22 in the wingtip of CZP and carried an over/under .22 hornet/.410 turkey gun in TWN, but I think Merlyn didn't like to think too much about survival kits. He focused on getting to his destination and not making it only half-way. He liked Jean's cooking, and he counted on being home to eat it, no matter what the hour.

Sometimes, when the weather is bad, the best choice is to stay on the ground. Passengers are always the worst judge of weather. An important engagement or appointment "clouds" their judgment and they can put a lot of pressure on a bush pilot to fly. Often the pilot will cave and take the trip, and the passengers only sometimes live to regret their pressure. It is much better to be on the ground wishing you were in the air than in the air wishing you were on the ground. Merlyn never pushed any of his pilots to fly in weather that made them uncomfortable. All the pilots I interviewed volunteered stories of times they told Merlyn they were not comfortable with the forecasted weather on a flight plan. Merlyn never questioned their judgment. He just took the load himself. There were a few times, especially as Merlyn aged that even he refused to fly.

The following account is about a time Merlyn at first chose not to fly. Many versions of this story have floated around bars and bush camps over the years. I heard the story from Don Stewart himself when he was fishing at Nonacho Lake. At the time he shared this story Stewart had consumed a few drinks, and he was also known as a man that could tell a good story so I cannot vouch for its absolute veracity, but like everything in the North, my inclination is to believe that, if anything, it is understated. McKay also remembered this incident and filled in some of the details.

Stewart, the mayor of Hay River and speaker of the Northwest Territory Legislature, wanted to get to Yellowknife for an important meeting. He had missed the sched, a daily milk run by a Pacific Western Boeing that connected the biggest settlements of the Western Arctic, so he called Carter Air Service to charter. The temperature was 52 below zero Fahrenheit and Merlyn, just that season, had decided not to fly anymore when the temperature is lower than 50 below. Too many things can go wrong and often do. The plane has to be warmed up, and then, when exposed to frigid weather, cracks can develop. Tempered steel will, at 50 below, become brittle. Plastic can tear like paper. Avionics can freeze up and fail. Ice fog can form in seconds. If you do go down at that temperature, it quickly becomes a survival situation. 52 below for a quick run to Yellowknife? It just didn't make sense and Merlyn refused his friend.

In those days, there was another air charter outfit in town. Stewart called them and in 45 minutes met the pilot at the airport. The plane was a Cessna 185 on hydraulic wheel skis. The wheels could be lowered beneath the skis for service on cleared tarmac or used on frozen lakes and rivers with the skis down and the wheels pulled up. The pilot had a tent over the front of the aircraft and the hose of a Herman-Nelson heater was pumping warm air into the engine compartment and cockpit. Herman-Nelson diesel heaters made much of northern winter aviation possible. They are dependable heaters that put out whopping numbers of BTUs. It was sunny at the airport and the pilot, Stewart and his assistant, were airborne in minutes. It was normally only an hour flight to Yellowknife. Stewart noticed the compass was spinning and mentioned it the pilot. The pilot shrugged and remarked that none of the avionics were working well either. His plan was to climb high and shoot straight across the big lake to Yellowknife.

There are a couple of buildings in Yellowknife that stick high above the horizon and can be seen 50 miles away as soon as you gain some altitude. Unfortunately, almost as soon as they started to climb, ice fog developed, and visibility dropped dramatically. The horizon disappeared as the white fog and the white surface of Great Slave Lake merged into one. Without a visual horizon or instruments on the control panel that create an artificial horizon, it is almost impossible to keep the wings level and maintain controlled flight. Without proper instrumentation and training, a pilot will soon begin to "roller coaster" by increasingly overcompensating and within seconds the plane will go into a dive or a spin.

Stewart's pilot cut power and went into a quick descent to keep eye contact with the ground. Ice fog is common when the temperature gets very cold. The air is simply too cold to hold any moisture; even though the air is extremely dry, it is "saturated." Near towns, habitation fog results from the condensing vapor of escaping warm air, exhaust or smoke. On cold days, the fog from a cabin's stove pipe can be seen for miles, and idling cars in front of the Hay River Hudson Bay Company store could create their own fog bank. But ice fog can also happen in nature, and because the dew point in frigid air is always close to the ambient temperature, it can happen fast, and it can extend right down to the ground. The crystals in the ice fog create sundogs, amber-tinted rainbow balls, and a rainbow wreath around the sun.

On this day, the ice fog was extensive, and the pilot was eventually forced to land on Great Slave Lake to wait for the conditions to improve. The temperature was now 58 below zero. Both Stewart and his assistant were second-guessing their shoe choices as the cold air quickly sawed through their oxfords. Even with the Cessna idling and the heater going full blast, it was cold. When it gets 50 below or colder, ambient (not wind-chill, wind-chill is baloney), the cold begins to creep in. The rivets in the plane's cabin were white, fuzzy buttons of frost. Stewart pulled up the hood of his parka; it crinkled.

An hour later the ice fog had lifted enough to continue the journey. But the plane had sat so long on the ice that the skis had frozen and the lake ice held the plane like flypaper. The pilot revved the engine and pumped the wheels down and up through the skis to try to break free, but without success. Finally, Stewart's assistant climbed out of the plane, and in the frigid prop wash of the 300-horsepower engine, was able to push the struts and break the skis free. The pilot goosed the throttle. The plane surged ahead. Much to Stewart's astonishment, the pilot continued to apply power, leaving Stewart's assistant farther and farther behind the aircraft.

"My God, what are you thinking? You cannot leave that man back there on the ice!"

The pilot applied left rudder and made a big sweeping turn on the snow. In the distance, with arms flailing, the stranded passenger was barely visible, ghostlike in the ice fog. So desperate to escape the cold and with his adrenalin still pumping from being left behind, the guy nearly staggered into the spinning propellor in his urgency to reach the door. He climbed into the plane with eyebrows, coat, and pants white with frost, and his eyes bulged with terror. Dressed the way he was, to be left behind on that lake and lost even for an hour, would have been a death sentence.

Airborne, the ice fog was still a problem, and the pilot was forced to fly at an altitude of less than 100 feet above the surface of the lake. He flew west, and when he reached the shoreline, he decided to ignore all his avionics and simply follow the coastline all the way to Yellowknife. The dark conifers provided some contrast and perspective to the view out the windscreen. The pilot was able to double his altitude and Stewart relaxed. Although following the coast would take a little longer and might mean missing the bulk of the meeting, Stewart was relieved to know the pilot was on course. The meeting no longer seemed so urgent.

After what seemed like an hour later, Stewart's assistant, who remained suspicious of the pilot, broke Stewart's relaxation by shouting, "That is the same clump of trees. We've been by this spot at least three times." With the low altitude and spinning compass, the pilot had been following a coastline. Unfortunately, it was the coastline of Big Island. Instead of progressing toward Yellowknife, they had repeatedly been traveling the perimeter of the island.

At least they knew where they were. Big Island is only about 40 miles from Hay River, but it's farther from Yellowknife than Hay River is. Fuel levels were becoming a concern. The pilot banked away from Big Island and first attempted to steer a course with his directional gyro. But the gyro had been sticky and unreliable the entire trip. With the compass spinning, there was no way to confirm a heading, and it was hard to trust the gyro over the feeling in the seat of your pants. The pilot began a wandering course, hopefully in the direction of Hay River. Attempts to fly more than 100 feet above the surface of the lake resulted in a loss of horizon. The featureless lake hypnotized everyone in the cockpit and required the pilot's constant vigilance to fly the plane straight and level. No one in the plane truly had any idea where they were. Finally, a dark spot appeared on the icy horizon. It first looked like a shrub, but their eyes had no reference of size. After a couple of blinks, the shrub had transformed into a commercial fishing camp. Without hesitation, the pilot reduced power and landed.

There were three white canvas wall tents, a half-dozen fuel drums, and a Bombardier. As the plane taxied up to the camp two fishermen emerged from the biggest tent. A small, wiry man, wearing a wool anorak trimmed with a strip of wolverine fur and a much larger man dressed in a canvas snowmobile suit walked toward the plane.

The pilot pushed open the door against the prop wash of the idling airplane.

"Too cold for flying, boy. Sixty below zero. Too cold for fishing, boy!" The short man, Eugene McKay, shouted with a twinkle in his eye and a big grin. "We were just getting ready to go into town. Boys! Too cold today."

"How far is it to Hay River?" the pilot asked.

"Maybe 40 miles."

Stewart and his partner had gotten out of the plane, and the younger man muttered to Stewart that there was no f------ way he was getting back into that plane or ever flying with this pilot again. The white tent was stuffy and filled with tobacco smoke and the smells of hot lard, frying fish and gasoline. But it was warm. A fuel oil heater large enough to heat a small house shimmered beneath heat waves. The plane passengers welcomed the blast of heat. They had been chilled since leaving Hay River. Even so, frost was fighting the oil heater. The makeshift plywood floor was caked with packed snow and the lowest walls of the tent looked like the interior of a freezer in need of defrosting. Three other fishermen lounged on a bench smoking cigarettes and watching. These fishermen had worked Great Slave Lake for years coming north first to work for George Carter. They were all Cree from northern Saskatchewan. Although McKay could not read or write, he managed the camp with the aplomb of an MBA, and what's more, he knew the lineage of all the men who worked with him back a half-dozen generations.

McKay sold the pilot 20 gallons of Amber fuel. With the plane still running, a wiry boy of 15 with a cigarette hanging from the corner of his mouth rolled a 55-gallon drum close to the plane, righted the drum and helped the pilot pump in the fuel. Then they both retreated to the tent to warm their hands. There was no big danger of fire in that cold air. At 50 below, you can throw a lit match into a pail of gasoline and the match will go out. But a cigarette may have been hanging out of this boy's mouth when he was refueling in the middle of July. Life jackets, fire extinguishers and the bother of ever unloading a gun are all foreign to the Cree.

Stewart began negotiations at once for a bug ride back to Hay River. The fishermen had planned to leave soon and close the camp until the weather warmed. With themselves and a half load of fish, the Bombardier was already overloaded. But a change of plans comes easy to these adaptable people, and after a few minutes of discussion, it was decided that McKay could take the fish and the two passengers to Hay River and return to the camp before dark to pick up the rest of the crew. The pilot could fly back or wherever alone.

The adventure was not quite complete. A bug ride on Great Slave Lake at 60 below zero is not undertaken without trepidation. Stewart and his partner were taking their time, sipping coffee and celebrating being back on the ground. They had missed their Yellowknife meeting and were in no rush to share a 90-minute bug journey to town with 500 pounds of whitefish.

McKay also knew at this extreme temperature that there would be a wide crack of open water and it was going to take some speed to "jump the crack." To Stewart and his assistant given the choice of climbing into the bug or into that plane, the bug sounded good. The earlier urgency of the Yellowknife meeting seemed silly. Stewart and his partner were eager to see their families again but were in no big rush to leave the relative comfort of the wall tent.

As the final arrangements were being made, and over the hiss of the oil heater, Stewart heard the plane revving up. Stewart said, "What the hell is that pilot doing now?" But he didn't finish his thought, because on the other side of the tent, still covered with frost, stood the pilot. Stewart pulled back the sash of the tent, and there was the empty plane. The engine was starting to race, and the tail wheel was already up off the ground in flying position. The skis were frozen to the lake and held the plane. Amber gas, a fine fuel for bugs, is too hot for aircraft. Without any load, the plane was eager. With their heads poking out of the tent's flap, everyone watched in horror as the starboard ski shuddered and broke free of the ice and the plane began to rotate. Just as the prop faced the tent, the other ski gave way and the plane began to taxi and build up speed, moving directly toward the tent. When Stewart dove for cover, the fishermen, who a moment earlier seemed in a bit of a stupor, were already under the table. Fortunately, the plane's elevator hit a fuel barrel, and like a pinball, the plane changed course and headed off disappearing into the ice fog of Great Slave Lake, happily missing all three tents. The pilot implored McKay to fire up the bug and chase it, but McKay just smiled," I don't think so, boy."

Stewart, his assistant, and the dejected pilot had not eaten for several hours, so Eugene was going to cook up some whitefish and bannock before they began the journey to Hay River in the bug. The meal was almost ready when the clunky and friendly sound of Merlyn's Single Otter rumbled through the still air. Merlyn banked around the fish camp on his approach. Stewart's assistant, who was starting to come unglued, had removed his topcoat and was frantically waving it as he danced in front of the tents screaming for rescue.

Merlyn had become concerned when Yellowknife Air-Radio reported the Cessna overdue. Then when the runaway Cessna had flipped after hitting a pressure ridge about seven miles east, it had triggered the planes emergency locator transmitter (ELT). Merlyn had found the empty plane and followed its ski tracks back to the fish camp. It might have been too cold to charter, but when there was the possibility of someone in danger, Merlyn did not hesitate to fly - not even in 60 below cold. A half hour later, the three adventurers were happily back in Hay River.

There are times in life to break the rules, even your own rules.

The Cessna pilot sent Stewart a bill for the flying a few weeks later. Stewart did not pay. The crashed plane was salvaged, but the charter business moved on after the following summer season and left Carter as the only bush pilot service in Hay River.

Some of the pilots who once flew for the Carters recounted stories of other charter owners that fired pilots on the spot because they refused to fly in bad weather, or because they flew an extra circuit to check the landing zone, thereby "wasting fuel." Merlyn and Jean did not do that. They expected an honest day's work out of their pilots but treated them with respect almost as if they were family. The Carters also made sure their pilots had what they needed. When

Clay Gamble complained that he did not want to fly an airplane with a defective artificial horizon, Merlyn ordered one expedited from Edmonton and had it installed within a day. Gamble went on to be a captain for Air Canada flying Boeing 777 and credits the Carters for getting his career started. Larry Buckmaster had a flying service in Hay River which included a Cessna 337 push-pull retractable. When one of his pilots neglected to deploy the landing gear and landed on the belly, sparks flying, props bending, Buckmaster asked Merlyn what he should do with the pilot. Merlyn said, "Well he probably will never forget to put down the gear again." Buckmaster agreed, but fired the pilot anyway, "I just cannot look at him without getting angry."

Water is the enemy of the aviator. Humidity is invisible water vapor in the air. Clouds form when the air is so humid it becomes saturated and cannot hold any more water vapor. It then condenses into clouds. Clouds cause turbulent air, icing and navigation problems for pilots. Cloud formation can happen fast. Fronts passing through, air cooling as it rises, and in the far north, in times of severe cold, a slight further chilling can cause ice fog to develop very quickly as it did on Stewart's memorable flight.

As his planes became technologically more sophisticated, Merlyn brought their instrumentation up-to-date. Readers might believe that if Merlyn encountered low ceilings and obscured horizons in winter, he could just set down on a lake and wait for the weather to lift. His planes were equipped with skis after all. This is partially true, and Merlyn had an uncanny memory for the location of every trapper's cabin, every old townsite, every fish shack, and shelter in the Territories. But landing on an unknown lake on a cloudy winter day can be difficult. He learned from his first crash with the 170B, that snowdrifts and ice ridges can be difficult to discern and can flip an airplane like the bail on a mousetrap.

Merlyn was a master at flying overloads and flying weather. He seemed to know every fence post and rock surrounding Hay River. Once, coming into Hay River from Nonacho on CZP he was notified by Flight Services that the Hay River airport was closed due to low ceilings and zero visibility. Even the Pacific Western jet was held waiting on the tarmac for the weather to improve before being cleared for take-off. When Merlyn radioed the tower, he was told the airport was closed, and unless he was low on fuel and wanted to declare an emergency he would have to divert to Yellowknife. Merlyn radioed in, "I am low on fuel and I declare an emergency." He landed successfully on the cross runway. The visibility was so bad the tower could not even see if he came in. But he did, dead center on the gravel strip.

Clouds can also cause problems in summer. Low clouds can not only impede navigation, but they can also force pilots to fly so low that even moderately high points can become obstacles to flight. Radio towers, power lines, even hills can be struck with disastrous results. Merlyn and his pilots avoided those type of collisions. There were times when clouds were so low and visibility so restricted that Flight Services closed, not only the Hay River airport, but even the river to aircraft take-offs and landings. In those situations, Merlyn would radio the tower and say he would land on the lake and taxi into the Hay River and to the float base. As many times, as Hay River Airport was closed, no one ever actually saw him land on the lake. Merlyn knew where the river was even when he couldn't see it.

Merlyn was a master at pilotage and more than that he became a master of pilotage without even a map. His son Myles remembers Merlyn flying a group of sport fishermen over 200 miles to a pristine lake on the barrens. When one of the fishermen asked Myles for a map to look at, Myles asked his dad, and his dad shook his head; just pilotage from memory.

In flight school, 50 years ago, much more time was spent on training pilots to be skilled at dead reckoning. Even Merlyn's earliest planes were equipped with a compass, and a Directional Gyro (DG) and these were useful for flying a course based on dead reckoning. For example, if a destination was 100 nautical miles due west. A pilot could fly a heading of 270 degrees for exactly one hour at 100 knots and he would reach his destination. However, for dead reckoning to be accurate pilots need to be cognizant of winds aloft and figure those variables into the plane's heading. Winds aloft were not reliably known or forecast in the early years and could not be accurately detected and accounted for while flying. In strong crosswinds, strictly following a compass heading can substantially alter the plane's route. A headwind can also delay the estimated time of arrival (ETA). If the destination was a trapper's cabin or a mining shack, flying a heading alone would not be nearly accurate enough. Even if the destination is a town, flying a heading can be tricky. I can remember flying a compass heading on XXO to the town of Eskimo Point. From my calculations, I was overdue but could see no sign of the town below. I did not realize that the topography near the Hudson Bay is so flat that when the Bay is frozen it is near impossible to ascertain where the coast is. In exasperation, I tried to reach the aerodrome on my VHF radio. After a couple of attempts with only static as a reply, a crisp voice with a British accent answered my call. It was a BOAC commercial pilot flying a great circle route over the pole. He had me on his radar and was directly over me just a few miles higher. He informed me that I was on course for Eskimo Point, just 50 more miles. He advised me to descend a bit as unbeknownst to me I was bucking a 40-knot headwind.

Northern pilots by definition flew at high latitudes and the closer a compass is to the magnetic pole the more it has a tendency to dip or sluggishly drift. Mineral deposits on the ground, iconic of the Canadian Shield, can also distort compass headings, and rough air can cause compasses to spin. A directional gyro (DG) was a key instrument on the panel of even on Merlyn's earliest planes. The old DGs required a vacuum that was obtained by a venturi tube mounted on the fuselage. These venturi DG's required frequent compass resets, were subject to icing and although a major navigation aid, were not completely reliable. Although useful, especially if the pilot is turned around, dead reckoning skills are a distant second to pilotage for bush pilots trying to find their destinations and to get safely back home. Global Positioning Systems (GPS) have changed all that and northern pilots now can travel point to point with the simplest of navigation.

Not only could clouds create visibility challenges; not only was there the threat of hitting something; there was also the chance that the clouds could hit you. Once flying the Single Otter IOF back from

Nonacho Lake with his family, Merlyn encountered severe summer storms. As he weaved the plane around the thunderheads, lightning bolts flashed on both sides of the aircraft.

Merlyn's grandson Jason asked, "Grandpa what will happen if we get struck by lightning?"

Merlyn replied, "Some of us will go to Heaven, some of us will go to Hell"

Humidity can cause other problems. It may seem like a paradox but saturated air is less dense than dry air. That is one reason why thunderheads build upwards. For pilots, the less dense humid air means longer take-offs, lighter loads and slower climbs. On a hot, humid, dead calm, day it is a big challenge for a float pilot to get airborne and sometimes stay in the air. This is one reason Merlyn liked the Cessna 180 so much. He often said, "If you can get that plane off the water it won't come back down." he would add with a bit of a sheepish grin, "I cannot say that about a Single Otter."

Once on a run to Nonacho and less than 50 miles out of Hay River he was forced to land his IOF on Polar Lake when he hit some hot humid air and was unable to maintain altitude even at full power. In April, with the Single Otter IOF on wheel-skis, with family aboard, skidoos and more, he got off fine. As he started to climb and as the plane lost the ground effect (the sandwiched air between the wings and the ground) the plane came back down hard on the runway near its end. The story is, he unloaded a case of toilet paper and tried again. He got up and stayed up that second time. There was something about Merlyn that every plane needed to be loaded to capacity. Once flying with him in the 180 TWN on floats we had full tanks, three people, 200 pounds of gear and a 90-pound propane cylinder (a load and a half). I remember a three-mile run on the river, Merlyn nursing the plane up onto the step and rolling on one float to finally get off. But as it was a Cessna 180, once you were up, you were up, but barely. We couldn't climb much that day until we burned off fuel flying the first 100 miles over Great Slave Lake. I had a good view of small pieces of driftwood and it would not have been much of exaggeration to say I could have scooped a cup of water reaching out the window. Merlyn trained his pilots to carry big loads and his nephew Stewart Portier and his sons became so adept at flying overloads that when they took jobs with other companies in the south they did double-takes at how small the loads were that they were expected to carry.

Liquid water and clouds are a big enough challenge, but when water freezes, then pilots have real problems. Frost on the wings can destroy lift. In damp winter weather, even modern jetliners are de-iced with antifreeze cannons before departure. In the early days of bush aviation this type of icing was not completely understood. Wing ice was the demise of Carter Fisheries earliest fish hauling workhorse, the Avro Anson. In 1961 Bobby Maclean attempted a takeoff from the West Channel ice runway, but there was just enough frost on the wings to destroy lift and he went into the willows. After ten years of hauling slimy fish, the plywood structure of that airplane had lost its integrity, "Nails just slid out from their holes" so no effort was made to salvage or repair it. Nothing lasts forever. After that lesson, Merlyn's pilots removed wing ice by rubbing a rope like giant dental floss, back and forth over the wings and elevator.

Chuck McAvoy never learned this lesson. Perry Linton watched McAvoy take off in a Stinson from Pine Point to Yellowknife with snow four inches deep on both wings. McAvoy nonchalantly believed, "It will slide off on take-off." It did, but only on the left side, causing the plane to make a violent yaw only a few feet above the ground. McAvoy had great "seat of his pants" flying skills and was able to horse the airplane through the trees and out over Great Slave Lake where the rest of the snow fell off. Then it was just 70 miles of open water, in a single engine plane, direct to Yellowknife.

Even if a pilot de-snows the plane with a broom and de-ices the wings with rope, he is not necessarily exempt from problems with ice. Ice can form on the wings, windshield, pitot tubes, venturi tube, even on the spinning prop while flying. What happens when ice forms on the wings in flight? When flying in damp near freezing conditions, especially in clouds, ice can build up fast and disrupt the airflow over the wings and add weight to the payload. If the plane is already overloaded, the pilot may have only seconds to find a different altitude where icing is not occurring and where the ice will melt off before causing the plane to stall and spin. Some planes could "carry ice" better than others, but having big popsicles for wings is never good. The pitot tube on the outside of the plane senses the plane's airspeed. Its tiny air hole is one of the first surfaces to freeze over. When that happens, the airspeed indicator drops to zero further complicating flying decisions at a time when it's critical for the pilot to maintain airspeed. Heated pitot tubes were standard on Merlyn's later planes to avoid that problem.

Ice was not only a problem on the outside of the plane but in the engine as well. Today's piston airplanes have a fuel injection system, but all the old planes used carburetors. When icing conditions were not present on the wings, carb ice could still be a problem. Carb ice forms when humid air coming through the venturi of the carburetor cools and condenses as ice. Left untended the ice will build up and starve the carburetor of oxygen causing the engine to flood and stall. Even in mid-summer, especially on the fringe of cloud, pilots would periodically pull a cabled controlled vent open that would apply heat gathered by a muff around the exhaust manifold and routed to the carburetor. Pilots would periodically leave carb heat on for a few moments to melt any ice that might be forming.

Water in the fuel itself could be a minor problem in the summer but a major problem in the winter. In above freezing weather a little water in the fuel will just cough through the engine but a serious amount of water can stall the engine and make restarting difficult. Water could get into the fuel by condensation in the fuel tank the same way it can get into an automobile gas tank. The location of fuel tanks in many planes is at the wing root and the sun on the wings makes condensation more likely than in a car. Water is heavier than avgas so it settles at the bottom of the tanks. Pilots usually drain a small amount of fuel from each of the wing tanks' sumps through shuttlecocks to check for water. The Single Otter has its fuel tanks in the belly of the plane which makes condensation less likely (but crashes more explosive). Water can also get into the avgas when pilots refuel from drums. The big barrels can leak or, if not full and sealed, are subject to condensation. To guard against putting water in the fuel tanks from the drums, Merlyn would filter avgas pumped from the drums by passing the fuel through a funnel, lined with a thick felt hat. Fuel would pass through the hat but water and sediments would remain in the hat.

Below freezing, water in the fuel can have very serious consequences and pilots often add alcohol to their fuel to prevent ice from forming. Ice can plug a narrow fuel line and starve the carburetor of fuel. This can cause the engine to stall and sometimes make it impossible to restart.

I don't know if icing made Merlyn nervous when he was younger, but by the time I knew him, nothing seemed to fluster him. Passengers would share memories of Merlyn asking them to take control and fly a heading while he took a snooze. What was always seared in these "co-pilots" memory was when a fuel tank would run dry and the engine would cough and quit. Nothing is quite as quiet as a Single Otter when the engine quits. Merlyn, always awakened by the silence or by the yells of the passengers, would nonchalantly switch tanks and restart the engine. He said he liked to run the tanks dry occasionally to clean them out, but not while Jean was in the plane.

Docked on the Hay River. Merlyn's Cessna 185
(courtesy of Carter family album)

Cessna 185
Data from Cessna
General characteristics
•Crew: one
•Capacity: five passengers
•Length: 25 ft 9 in (7.85 m)
•Wingspan: 35 ft 10 in (10.92 m)
•Height: 7 ft 9 in (2.36 m)
•Wing area: 174 sq ft (16.2 m^2)
•Empty weight: 1,748 lb (793 kg) 1,910 lb on floats
•Gross weight: 3,350 lb (1,520 kg)
•Powerplant: 1 × Continental IO-520-D , 300 hp (220 kW)
•Propellors: 2-bladed constant speed, 6 ft 10 in (2.08 m) diameter
Performance
•Maximum speed: 155 kn (178 mph; 287 km/h)
•Cruise speed: 145 kn (167 mph; 269 km/h)
•Stall speed: 49 kn (56 mph; 91 km/h)
•Range: 720 nmi (829 mi; 1,333 km)
•Service ceiling: 17,150 ft (5,230 m)
•Rate of climb: 1,010 ft/min (5.1 m/s) 960 fps on floats

Chapter 12
Egg Island

Egg Island on Great Slave Lake got its name from gull eggs that were harvested here by the indigenous people Samuel Hearne wrote about the Chipewyan people eating gulls' eggs during his journey to the Arctic Ocean in 1763. At that time, the most esteemed eggs were those that contained developing chicks. The people would eat these raw and spit out a bundle of emerging feathers. One hundred and fifty years later the culture had changed, and eggs collected on Egg Island were put in a bucket. Eggs which floated we're discarded as being too developed. The heavier eggs were used in cooking. By the time I worked in Fort Resolution almost all the eggs people ate were chicken eggs from the south and bought at the Hudson Bay Company store. Times change cultural norms. It's not that one way of judging what is fit to eat or fit to do is better than another; it is just our thinking that makes it so. Egg Island is in this narrative because that is near where Merlyn made a forced landing in 1999.

Merlyn's marriage to Jean was a passionate romance. The days of "hide and seek" continued to the end. Both were devoted to each other and there was nothing Merlyn enjoyed more than an evening in his Vale Island home with his family, having a glass of wine, playing some guitar, while Jean whirled around the kitchen. If Myles was visiting, it would always be, "Myles, play that tune on the piano again." When the guests went home, it was with Merlyn that Jean could drop all pretense, curl up in his strong arms and recharge from the trials and tribulations of family, business, and life.

There were some stormy days too. In the late 1960s Jean had become a devout Jehovah's Witness or as she says, "taken the truth." Merlyn was one of the least judgmental men I have ever known, and he respected Jean's beliefs. Respect and acceptance does not mean agreement, and Jean's strong beliefs did lead to conflicts and hard feelings. Jean was profoundly saddened that Merlyn would not embrace the Jehovah Witness faith and although I do not believe it diminished their love for one another, it was certainly a schism in their relationship. Although Merlyn preferred to avoid even talking about it, he would confide to me that it was a source of friction in their marriage. It was also a flashpoint in their family. Dean strongly adopting his mom's religion, Kandee hanging with her dad and Myles bouncing back and forth. It was even more contentious when Merlyn's parents were alive, especially regarding their grandchildren and especially at holidays. Christmas was a challenging time for the Carter family. Jean would not permit a Christmas tree in the house, decorations, or a Christmas present exchange. Merlyn reluctantly accepted Jean's wishes, but there was some brooding and sniping in December. I remember one holiday season when Merlyn with a twinkle in his eye and an askew glance at his wife teased that the six-foot evergreen houseplant in the dining room was their Christmas tree. The next day I noticed the feisty Jean had pruned it to a stalk. Despite the heartaches and a few flashes of anger, forgiveness, a jovial focus on the now, and enduring love pulled them through. After Merlyn's parents passed away and the kids grew up, some of the friction over their religious differences cooled and smoothed. Still, the holiday season was an unsettling time for Merlyn.

Back to Egg Island. On Christmas Day of 1990 Merlyn decided to get out of town and fly out to the Simpson Islands to check in on a commercial fishing crew and bring them some Christmas cheer. He would take his Cessna 140 YNC. Merlyn had bought this 35-year-old plane in 1984. There may have been a bit of nostalgia with the purchase. A 140 was his dad's first plane. A more likely explanation was that the 140 was just a cheap plane to operate. The 85-horsepower engine sipped fuel and the tail wheel configuration made it a sweet little bush plane. I think the real answer to the purchase was, the 140 could realize a dream Merlyn harbored for a long time.

Three decades earlier Merlyn had chosen Nonacho Lake for his sport fishing camp for three reasons. It had the best lake trout fishing of any of the dozens of lakes he tested when pioneering for commercial fishing prospects. Most of the lakes in the region contained some big trout. What distinguished Nonacho was how great fishing sustained itself year after year after year. Even after 30 seasons of sport fishing, trophy lake trout were still being regularly caught off the point, less than a mile from the camp. Nonacho is also one of the most beautiful lakes in the world. To step off the float of a plane and onto a Nonacho beach, it is almost impossible not to suck in a mouthful of air and look 360 degrees in wonderment. Dozens of islands - some ringed with blond sand, dark brooding groves of spruce, towering granite cliffs etched with black lichen, ribbons of sandy beaches, steep-sided eskers crowned with tundra meadows and studded with jackpine it is all right there. When the north wind is gently blowing the air is so clear it tricks your vision and you feel as if you are on the moon, no longer shrouded by atmosphere. The Carter Camp itself was positioned on a crescent of sand, almost devoid of bugs and protected from cold northeast gales. Nonacho was off the beaten track. Of course, the entire Northwest Territories is off the beaten track. But Nonacho even lacked the scars from winter ice roads and played-out mines that dot an area of similar in beauty north of Yellowknife.

Big trout, sustainable fishing, astonishing beauty, the one missing piece was an airstrip. Every sheet of plywood, every aluminum fishing boat, every diesel generator, and most importantly every drop of fuel had to be flown to Nonacho on floats or on skis. Every fisherman had to be transported by float plane. This was not a big issue in the early years. Time changes things. The price of gasoline had skyrocketed. The 230-mile flight from Hay River burned a lot of fuel, especially in a big float plane. The potential clients were becoming more persnickety. In the sixties, a tent frame, a Coleman stove, a little home-made wooden boat with a 6-hp kicker was all that was expected. Later, tourists started to expect ice, cast iron wood stoves, propane, cabins, electricity, aluminum boats with big electric start equipped outboard motors, depth finders, showers, more and more. Biggering begets more biggering. Merlyn flew out refrigerators, ice machines, generators, bigger outboards, snowmobiles, ATVs, and tractors. For the most part, Merlyn enjoyed the growth of the camp, his favorite thing, was the little store's air conditioner. He also liked the challenge of external loads, whether it was tying big boats to the floats or the thought of transporting a piano with half the instrument sticking out the cabin door. At the end of the twentieth century it was likely the busiest sport fishing camp in the North fully built and supplied by bush planes on float and skis. No barges, no wheeled planes no remnant buildings from a played out mine. Early on there was a single attempt to bring a Bombardier to Nonacho. It was a one-way trip and sits idle on an isthmus in sight of the camp. An airstrip would have made transport easier and more economical.

Flying wheeled aircraft direct to Nonacho would have also helped in marketing. The pace of life has speeded up. In the sixties, tourists had time to fly to Edmonton and catch the northern mainliner to Hay River. They had the time to spend a couple nights on the way and then a week at Nonacho and a few days getting home. Increasingly, tourists could not afford the time. Merlyn dreamed of an airstrip at Nonacho. With his big DC-3 he could fly passengers direct from Edmonton or northern Saskatchewan to his camp.

My cabin was on an esker just a mile from the Carter camp. Merlyn was convinced he could build an airstrip on the esker near my cabin. I couldn't see it, at least not an airstrip big enough for a DC-3, but the worm was gnawing at Merlyn's brain. I think the biggest reason for the 140 was so he could realize his dream of landing a plane on the ground at Nonacho, even if it was only a two-seater. Merlyn eventually gave up on the esker strip, but one April on the ice, he did drive his new Ford loader over from the island and started clearing a runway on a rocky beach about three miles from his camp. I am unsure if he ever landed on that strip: it was short, perpendicular to the prevailing wind and rocky, but I have no doubt he could have landed the 140 there. In the Pine Point staking days, it was said that Merlyn could land his Helio Courier on the shell of a turtle (had there been any turtles in the NWT). With a stall speed of close to 40 mph, the Cessna 140 could also come in low and slow. Merlyn never realized his dream of flying his customers direct to Nonacho from the south. The closest he came was an arrangement to bring a load of Minnesotans in from Battleford, Saskatchewan, in the DC-3 for some April ice fishing. Unfortunately, early warm weather made Merlyn second guess the prospect of landing the wheeled DC-3 on a slushy lake, so he ended up bringing the crew to Hay River first, and then flying them out to Nonacho on his wheel/ski equipped Single Otters.

Back to Egg Island. The route to the Simpson Islands would take Merlyn across the frozen Great Slave Lake, just northwest of Fort Resolution directly over Egg Island. Fish camps were usually alcohol-free, and Merlyn would not have been bringing booze, but he might have a box of Christmas Mandarin oranges or some sweets for the boys. Odds are there was not much else in the plane. The YNC with full tanks has only about a 350 lb payload and Merlyn's weight would have taken up most of that. The 140 like all planes of that era had a weak heater. It took cabin heat off a muff on the manifold, and with an air-cooled 85 hp engine on a cold December day, not much heat was coming off the manifold. In practice, pilots and passengers of a 140 would dress for the outside ambient temperature. Merlyn was wearing a snowmobile suit with heavy boots, wool hat and mitts. About an hour out from Hay River the 140's engine quit. Likely similar to the cabin heat there was not even enough heat coming off the manifold for the carb heat vent to clear accumulating ice in the carburetor and the engine starved. Merlyn, on one of the shortest days of the year, was without power and a thousand feet above Great Slave Lake. Early in the winter, there was little snow on the lake and what there was had been wind packed. Still, a powerless (dead stick) landing is challenging especially on rough terrain. A pilot prefers to have a little power to slow the descent so he can fly onto the lake. Without power, a plane is descending faster and the pilot needs to keep the nose up flying slow, just above stall speed, tail down. Merlyn landed the plane perfectly.

The situation quickly shifted from an aviation emergency to a survival situation. Merlyn made a few attempts to restart the plane with the battery and by propping (spinning the propellor by hand). He had no way to apply heat to thaw the ice in the carburetor. The days when pilots carried blow pots and canvas engine tents had passed. Anyway, with Merlyn in that little taildragger there was not much room for gear. Being miles from shore there was no fuel to stoke a fire. His only radio was a line-of-sight VHF and although only 14 miles from Fort Resolution, the aerodrome was uncontrolled and nobody monitored the radio, especially on Christmas day.

He could see the runway light of the Fort Res airport 14 miles away, but it likely looked closer. The temperature in Hay River was around zero Fahrenheit. He was not expected back in Hay River until dark. A rescue mission might not even be initiated until morning. Dressed for winter, and sitting around in sub-zero cold for upwards of twelve hours are two different things. Freezing to death was a real possibility. More than that, to Merlyn the thought of sitting in a cold cabin and waiting to be rescued was anathema. He tied up his bootlaces and started walking to Fort Res. The flat lake with just a thin coating of blown snow was easy walking, but 14 miles are 14 miles. Merlyn lived only a quarter mile from his office at the float base on the Hay River. In 30 years, he never walked to work. Merlyn could single handily load a 55-gallon drum of avgas and had a vise-like handshake, but now in his sixties, endurance fitness was not in his wheelhouse. In the pitch darkness of 6:00 pm, exhausted, and only by force of will, did he make it to within 100 feet of the airport runway. I know that runway in winter. The bank of the lake and brush act like a snow fence and deep drifts build up along the coast. Over the last hundred feet, Merlyn had to post hole through four-foot drifts. Each step took a Herculean effort. When he staggered into the RCMP housing compound located just behind the airport, he was near death. Before being rushed 100 miles to the Hay River hospital, he found the strength to apologize for interrupting their Christmas feast.

When a piece of machinery let Merlyn down whether it was a propane grill, a generator, a pick-up truck or a wristwatch, there was a good chance he would just trash it and buy something new. After his Egg Island ordeal, he sold the 140.

Merlyn would only purchase one more airplane in his lifetime, a Cessna Caravan. This was as opposite from the 140 as can be imagined. Fitted with amphibious floats it could take off from a runway and land on a lake. It was a high-tech airplane that could carry a Single Otter load but do it a lot faster and quieter. Merlyn did like the Caravan, but not too much. It was an expensive airplane to maintain and repairs went far beyond tinkering with a pair of pliers and a screwdriver. Nose wheel equipped and not a STOL aircraft this was not a plane for short bumpy runways.

Even when your heart is in them, not every dream comes true. There was never to be an airstrip at Nonacho Lake.

Myles Carter and Jason Froese in front of Merlyn's Cessna Caravan (courtesy of Carter family album)

Cessna 208 Caravan
Data from Cessna
General characteristics
•Crew: one or two
•Capacity: nine passengers
•Length: 37 ft 7 in (11.46 m)
•Wingspan: 52 ft 1 in (15.88 m)
•Height: 14 ft 11 in (4.55 m)
•Wing area: 279 sq ft (25.9 m2)
•Aspect ratio: 9.702
•Empty weight: 4,730 lb (2,145 kg)
•Gross weight: 8,000 lb (3,629 kg)
•Fuel capacity: 2,224 lb (1,009 kg/332 gal/1,257 l)
•Powerplant: Pratt & Whitney PT6A-114A turboprop, 675 shp (503 kW)

Performance
•Cruise speed: 214 mph; 186 kn (344 km/h)
•Stall speed: 70 mph; 61 kn (113 km/h)
•Range: 1,232 mi; 1,070 nmi (1,982 km)
•Service ceiling: 25,000 ft (7,600 m)
•Rate of climb: 1,234 ft/min (6.27 m/s)
•Wing loading: 28.674 lb/sq ft (140.00 kg/m2)

Chapter 13
The End of an Era

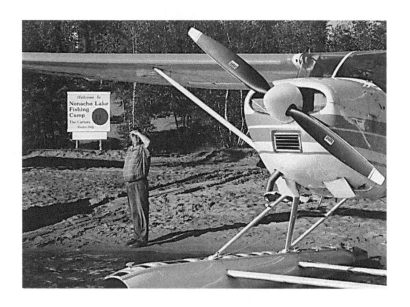

(photo courtesy of Carter family album)

On June 15, 2005, Merlyn Carter was killed by a bear. The unprovoked attack occurred on the shores of Nonacho Lake, a few feet from the store of his sport fishing camp. Merlyn had traveled all over North America, owned a condo in Maui and a newly remodeled home in Hay River. His favorite spot in the world and his wife's favorite spot in the world remained always to be Nonacho Lake.

The day before his death, he had been flown out to the lake in the 180 TWN by his grandson, Jason, now a licensed pilot and a young man of 22. The ice had just come off the lake and the first tourists of the season were arriving at camp on the 17th. There was work to be done to get the camp ready for tourists, and Merlyn had planned to start the generator, open up the cabins, turn on the propane and get the water for the shower house and store operational. A few years earlier, Merlyn had flown in the newest diesel powerplant. It was housed in a shed half bunkered into the sandy esker to muffle the noise, and located about 100 feet inland from the store.

Although called "the store" it is the main cabin and headquarters of the fishing camp. There is a large kitchen/dining area, two small bedrooms and the actual store with a counter, shelves and even a wine cellar dug into the sand beneath the floor and accessed by a hatched door. The store is fully equipped with electric lights, two propane stoves and ovens, woodstove, an electric refrigerator, freezers, an ice machine, washer/dryer, and even an air-conditioner. It is quite a change from the Coleman stove and canvas tent that Merlyn first pitched on that site over a half-century ago. The store is the epicenter of the camp which consists of a half dozen plywood tourist cabins and several sheds and tent frames.

At age 71, alone, deep in the wilderness, Merlyn was enjoying puttering around his camp, repairing and building whatever was needed. A 30-30 Winchester carbine lay on the dining table of the store with a half-dozen cartridges beside it lined up like little soldiers. Forty feet away another rifle, a .303 Enfield was loaded and leaned against the inside wall of the outhouse.

Merlyn had talked with Jean the evening of the 14th on the old HF radio, and all was well. Myles was going to fly in with Jean the following day. They were planning to clean cabins and get the boats and outboard motors ready to go. But between 10:30 that evening and 2:35 the following afternoon a terrible event occurred that would send shock waves across the Territories. The actual sequence of events will never be known, but much can be pieced together. Before going to bed, Merlyn left the store, closing the kitchen door behind him. He was carrying a flashlight, which likely meant he was checking on the generator. At this latitude in mid-June a flashlight is not needed outdoors, but might be required in the powerplant shed. It is likely that to conserve fuel, Merlyn was turning off the generator for the night.

Bears are typically timid creatures, although in the Northwest Territories they are less so. Aggressiveness is particularly true for the bears that roam the environs of Nonacho Lake. Being so close to the treeline they live in an interface zone where the range of the black bear and the tundra grizzly bear overlap. The only challenge to their supremacy comes from another bear and is often settled with violence. These bears have likely lived their entire lives without seeing a human, and so they lack the fear of humans typical of bears that live in southern latitudes and that have been hunted for generations. There may be other reasons for peculiar bear behavior in the far north.

Tim Beahen, a Fort Smith canoeist and teacher, shared a story with me of a Nonacho Lake black bear. He was paddling with his wife down the seemingly endless north arm of Nonacho. A bear had started to follow the canoe along the shoreline. At first this was an interesting and prolonged opportunity to observe a species which usually only grants canoeists a glimpse. But after a while it got creepy. The bear continued to follow along on the shore matching the progress of the Beahens foot by foot. "It was as if we were being stalked." Finally, after three hours, the Beahens were thinking about camping, but the bear was still there. Tim decided to shoot a round with his .30-06 rifle over the head of the bear to scare him off. When he fired the gun, an astonishing thing happened. The bear dove into the lake and started swimming directly toward the Beahen's canoe. Showing more restraint than I might have, the Beahens turned their canoe and out-paddled the bear toward the middle of the lake. After a prolonged chase, the bear turned and headed back to shore. The Beahens continued paddling and eventually camped on the opposite side of the lake.

A half-million black bears live in the United States and southern Canada. Close to civilization, seldom is there an account of anything more aggressive than bears robbing bird feeders or knocking over garbage cans. Rare are bears with predatory instincts directed toward humans. Perhaps that is because bears with that type of instinct did not survive long in southern Canada or the States. The gene pool in the south for that type of behavior may have dried up. In the far north, predatory bears are not so unusual. Aboriginal and white campers never stray far from their firearms. Spring is also a season when bears are on edge. Recently emerging from hibernation to a land that has not come fully alive makes bears cranky and hungry. Firearms at the Nonacho camp which are typically stored out of sight once the fishing season begins were kept closer at hand during the spring.

On the evening of June 14 when sleet rattled against the windowpanes of the store and the wind whipped up whitecaps on Nonacho Lake, neither gun was close enough. Just a few feet from the stoop of the store, Merlyn, and a black bear were in a life and death struggle. For seventy-one years this pilot pioneer had eluded death. He had flipped his 170 on its back, crashed and burned the Lockheed, put an Otter through the ice on Great Slave Lake and another in the trees at Nahanni Butte. Prop seals have ruptured and covered his windscreen with oil. Propellors have fallen off in mid-flight. His planes have caught fire. The wings of his planes have, in mid-flight, become sheathed in ice. Merlyn had survived engine failures, flat tires, shifting loads. He'd flown firefighters into the fringe of wildfires and performed rescues, medivacs, and searches. Many times, he flew in weather that would have grounded any other pilot. He successfully landed on rocky beaches, wind-whipped lakes, rushing rivers, fog shrouded runways, frozen meadows, sandbars, gravel bars, thin ice, deep snow. He had been stalked by wolves and harassed by grizzlies. He carried prodigious loads inside and tied to the outside of his planes that other pilots thought were impossible. And he was tough, even at 71, He could still toss a 90-pound propane cylinder into the back of a Cessna 180 like it was a stick of cordwood.

Likely this black bear was just passing through at the exact moment Merlyn left the cabin. The Carter's small dog Gus was in the cabin and would have alerted Merlyn if the bear had been long in the vicinity. Perhaps the bear hit Merlyn from behind and knocked him to the ground possibly breaking Merlyn's back. In any case, Merlyn did not go quietly. To keep rodents out from under the store, a chicken wire fence had been stretched between the store's supporting posts and stapled in place. In the midst of the struggle, Merlyn had pulled this screening from the posts and twisted it into a cable.

Perhaps he did this to pull himself under the building for protection or more likely to use as a weapon to force between the jaws of his attacker. In any case, his valiant efforts were unsuccessful. Alone, over 90 miles from the next nearest human, but in a land he knew as well as anyone, and at a place he best personified, Merlyn died.

Full of excitement about the upcoming and 39th fishing season at Nonacho Lake, Jean and Myles arrived at the camp at 2:35 the following afternoon in TWN. The previous night's cold wet northeast wind had banked around to the southwest, and the sky was clearing. As Myles taxied the 180 to the dock, they expected to see Merlyn walking down from the store. It is customary whenever a floatplane arrives to meet the plane at the dock and to assist in mooring and unloading. When Merlyn did not come down to meet the plane on this afternoon, it created an anxious moment for Jean and Myles. Where was he? Jean was concerned, thinking Merlyn may have suffered a heart attack or hurt himself cutting firewood. Even before Myles had tied up the plane, she jumped off the pontoon and onto the dock and ran up toward the store to search for him.

Merlyn had built his camp on a sand beach. The store is set back about sixty feet from the lakeshore on a modest esker ridge. Jean jogged toward the store. She was halfway there when she was startled by a bear who was half hidden by a black spruce, but less than ten feet away and walking directly toward her. The bear's eyes were riveted on Jean, and the bear's shoulders and head swayed back and forth as it moved menacingly and deliberately toward her. Saliva ran from the bear's lower lip in a thin continuous stream. The bear's ears were flat against its head. Coming from town, Jean still had a cell phone in her pocket. She flung it at the bear as she swirled around and ran back toward the plane. The bear kept coming. The grand-daughter of Vater who had led his family with undaunted courage halfway across the continent in search of a home and freedom, Jean said aloud, "I am not going to let that bear pull me down from behind." With grit that had been tested many times since her birth back on that Saskatchewan homestead, this little fireball of a woman turned and faced down that bear. She pulled off her nylon jacket and twirled it around her head, and she kicked sand into the face of the bear. But it kept coming, crouched low and swiping at her with its paws and lunging to within inches of her torso while showing its teeth. Although this whole encounter had only taken a few seconds, but Jean couldn't help but wonder why her son had not yet come to the rescue. Myles was already on the way. He had heard Jean shout, and when he looked up from the tie-down ropes and saw the bear chasing his mother, he leapt into action. Myles, already the father of a teenager and an accomplished athlete, looks bigger than his 6'2" frame and he came straight at that bear with arms extended high in the air, and shouting louder than he had ever shouted before. But the bear would not even look at him and remained focused on Jean, coming closer and closer.

When Twin Otter aircraft bring fuel loads into camp, the Carters use an eight-foot, steel, 30-pound ramp that attaches to the door of the plane and allows them to slide 55-gallon barrels of fuel into the water or onto the dock without damaging the aluminum pontoons of the aircraft. Myles reached down for that ramp. Using it as a battering ram, he drove the ramp hard into the bear's snout. The impact sent Myles backward, and he would later remark that he felt as if he had driven that ramp into the stump of an oak. There was no "give" to that bear.

The heroic act with the ramp did distract the bear from Jean. The bear ambled down the beach, disappearing into some scrub birch. As quickly as he could, Myles pulled the survival carbine from the cabin of the plane. With a .22 Hornet barrel on top and a .410 shotgun bore on the bottom, this survival gun is designed for small game, not bear, but it was certainly better than nothing. Jean climbed into the plane, alarmed that Myles was going after the bear, but also realizing that with Merlyn's whereabouts and fate unknown, her son had no choice but to pursue the bear, and finish what the bear had started. Myles stepped up on top of an old freezer that was against the wall of a small storage shed at the foot of the dock. Then he climbed up on to the shed's roof. From that vantage point, he scanned the beach area, but could not see the bear. Yelling to his mom to watch for a stalking bear he came down from the roof and worked his way to the front door of the store like a soldier on a mission. The door was locked, but he broke through a plastic window with the butt of his gun and went in. He found the Winchester and grabbed 4 of the cartridges that were lined up on the table. Thinking of the peril his father might at that moment be facing he did not dare waste time to load the magazine, but just threw open the lever and pushed a single cartridge into the chamber and stuffed the other three cartridges into his pocket.

When he opened the back door of the store, he saw the bear. It was halfway down the path to the generator shed. Myles did not allow himself to recognize what the bear was standing on. He took aim and fired. The bear dashed behind some trees as Myles chambered another round. Merlyn had once told his son, "Once you shoot at a bear, keep shooting," and Myles did. His second shot was at a black square of fur in the middle of a thicket of spruce and birch. Myles fired his third shot as the bear crossed back over the trail to the generator. Myles fired the fourth and final round and hit the bear's chest squarely. With a shudder the bear collapsed, lifeless.

The object on the trail that Myles had refused to believe he saw was, of course, his father. Barely recognizable, long dead, half eaten, and dragged twelve feet down the trail was a man who was loved by many. As the sun shone on Nonacho and the adrenaline faded from the hearts of Merlyn's wife and Merlyn's youngest son, it was replaced by immitigable sorrow and a reluctant admission that an era had been brought to an end.

The memorial service for Merlyn Carter was one of the biggest events ever held in Hay River. The hockey arena was the only space large enough to hold the crowd. People came in from across the North. In his lifetime, Merlyn had been generous and given a helping hand to many beginning pilots who were now anxious to acknowledge their respect. For anyone who wanted to attend the service, Joe McBryan, the founder of Buffalo Airways offered free air transport from Yellowknife to Hay River and back again in a DC-3. Days earlier, and as soon as he had heard the news, Merlyn's oldest son, Dean, flew north from Vancouver. Thinking he would have to spend the night in Yellowknife, he was poignantly surprised when two pilots from Air Tindi met him at the gate and flew him home to Hay River that evening. But the legendary bush pilot of the North had touched more than the hearts of pilots. In a land with few roads, it is the planes that haul in food and supplies, take out the sick and injured and reconnect families and lovers. Many who flew with Merlyn would fly with none other. He had earned the trust and respect of people across the Territories.

I did one of three eulogies at the service. I gave it all I had, but I couldn't stop wishing I could do more. I asked Myles, "Let's get TWN up and buzz the town, let's buzz the town a couple times."

But Myles said that those days of buzzing Hay River are over. "People are more civilized now. We could get charged, and even arrested."

Ugh! In many ways Merlyn's passing was more than his death. It marked the end of an era.

In the weeks that followed Merlyn's memorial I wasn't sure whether I would ever return to the North. When Myles had called me from the beach on June 16, I felt wave after wave of sorrow, regret, and loss. I also felt an unmitigated urge for revenge. Those feelings persisted.

I had lost my best friend. This man who never had a harsh word for me and in whose presence I felt that all is right with the world and with me, was dead. Why could have I not been at his side? That thought reverberated in my mind like the burst from a machine gun. Together we could have fought off that bear. Had I been there Merlyn would still be alive. Everything would be as it was. I would have the opportunity to sit by him, to see that twinkle in his eyes, that licking of his lips. To listen to his stories again. To hear him sing. To fly with him again.

Memorial plaque Hay River
(photo by author)

I wanted revenge. I wanted to kill the next bear I encountered. I wanted to take a baseball bat and walk into the fudge shops of Ely, Minnesota. I wanted to smash the figurines of bears, tear down the T-shirts with corny bear sayings and scream insults at the store proprietor who sold this junk and tried to pass off murderous black bears as innocent and cute creatures of the forest. Most of all I wanted a bear's blood on my hands.

I was grouchy with my family and angry at the world. My dreams were steeped in bear charges. At restaurants, I sat with my back against the wall always ready for a bear charge, even in downtown Minneapolis. I bought a Smith & Wesson .38 caliber handgun, not so much for defense, but to kill the next bear I saw. My friends asked me, "Why a .38? Such a puny handgun for bear?"

I responded, "I don't want the bear to die quickly." I was serious. I even fantasized about killing a bear with my knife. Wrapped in his arms, I would drag my fingers down its sternum force my knife in deep, push upwards, and stir the oatmeal until the bear went limp. It was a recurring dream at night while I slept, and a haunting daydream.

My opportunity for revenge came almost a year later. I was driving north on my way to guide a group of canoeists down the Kopka River in northern Ontario. It was to be a 10-day whitewater trip on the southern fringe of the Wabakimi Wilderness. The road to the seaplane base in Armstrong was desolate. We had chartered a Single Otter which had undergone a million-dollar turbo modification. The conversion would have made Merlyn shake his head. Decades had passed since the Otter was proclaimed a powerful short-field performer. The bush plane industry evolved as pilots demanded more and more power. Modern aviation writers now describe the old piston 600 hp Otter as, "underpowered". Although the turbo conversion increased the thrust by nearly 50%, Merlyn would have known better. He knew the size of the engine was only one factor in getting a load airborne. But Merlyn was dead, and I was seeing red, blood red. We came around a sweeping curve of the gravel road, a curve that would lead to a seminal moment. On all fours and eating greens along the roadside was a bear.

"Stop the car!" I shouted.

Because of Canadian gun laws, I was unable to bring my revolver across the international border. I cared not. Pulling an axe from my friend Woody's pack I clambered out of the truck and rushed toward the bear. The bear ran for the nearest spruce and ambled up 20 feet, as high as the thinning spruce would support him. Undeterred, I swung the axe and began to chop down the tree. Looking up, I saw a bear trembling with fear. He was clutching the trunk of the spruce with his claws, looking downward with a look of terror and innocence at his attacker.

I dropped the axe. Inexplicably, a tear ran down my cheek. I cannot explain it, but all my animosity toward bears evaporated. The bear looked pathetic. In that instant, my fear of bears also dissolved.

Vengeance, holding a grudge no longer grips my spirit. I am no longer interested in killing bears. I have not since suffered even a quiver of fear in the presence of bears. Since that moment I have seen plenty of bears, mostly grizzlies, plenty close, all without feeling a smidgen of malice.

As a canoe guide in arctic Alaska, it's typical to have run-ins with a dozen grizzly bears every summer. One morning, on a canoe trip down the Kugorurok River I was standing in the river brushing my teeth when a big grizzly came over the hill. He walked past my tent, and I swear he looked over at the tent flap and saw the double-barreled, ten-gauge shotgun resting on my foam sleeping pad. Then with a sly grin, he looked back at me standing in the river armed only with a toothbrush. Maybe he was tiring of chum salmon and relished some red meat. For whatever reason, he kept walking deliberately toward me.

I was unperturbed, but when the grizzly closed within 20 feet, I lifted my whistle from its lanyard and blew it with gusto. I learned this defense from an old Inupiat that lives along the Noatak River. The grizzly was unperturbed by the whistle and continued to walk toward me. It wasn't until my companion heard my whistle and started running toward me blowing her whistle that the bear decided to stick with eating salmon and turn away. I continued to brush my teeth. It's not by accident that both the bear and I were "unperturbed". Bears remind me of people. Often timid, sometimes opportunistic, they are usually quick to back down and run. If they want to fight? Okay, bring it on.

This past summer I was leading a group of 10 down the Noatak River. We had stopped for a lunch break on a fine sandy beach. I spied a grizzly almost a mile away walking down the beach toward our lunch spot. I figured we had about 20 minutes, plenty of time to finish lunch, and I sent my co-guide to coax our campers to finish lunch and pack up. I toted the shotgun down to the beach, downstream of the picnic and upstream of the bear. He sensed my presence and veered off the shoreline and into the willow studded tundra.

Sometimes bears are stubborn. Camped beside an arctic river, and after a day of bear encounters, I was again brushing my teeth, only to see a bear standing on two legs and staring at me from 100 feet away. When it comes to bears, ice may flow in my veins, but I am not stupid. Cradling the sawed off 10-gauge shotgun in one arm, I turned and walked two steps toward my tent. Looking over my shoulder, I watched the bear drop on to all fours and walk two steps toward me. I turned around and walked two steps toward the bear. He turned around and walked two steps away and then faced me. It was a stare-down. Then he sat. His back against a rock, his two legs were outstretched like a teddy bear on a kid's bunk. I sat, my back propped against a rock, my legs outstretched. Then slowly, the bear's head swayed forward until his chin came to rest on his chest. He was sound asleep. I quietly stood up retreated to my tent and slept like a baby.

Some bears, like some people, are killers. But like people, they are not designed well for it. Merlyn's death struggle with a bear might have ended differently a few decades earlier. In fact, even at his advanced age many of his friends said that they would have bet on Merlyn one-on-one with a 180-pound black bear. But for want of a gun or even a knife that is not the way it played out, and a great man died alone and not without a fight.

I am not sure where Merlyn's family is with bears. I know it was difficult for Jean to return to Nonacho, a place she and Merlyn had shared so many happy memories, but she did. In August of 2016 the family spread Merlyn's ashes on the lake and placed a memorial in a concrete slab on a knobby hill overlooking Nonacho.

His oldest son, Dean, struggled mightily with his father's passing. Dean visited me in Minneapolis and I helped him with a display at a sport trade show. In honor of his father, the maker of floats for airplanes, invited him to a party at a bar in the warehouse district of Minneapolis. When we got to the party we parked Dean's rental car in a nearby private lot. The party was overcrowded, with a line snaking down the street. We decided to give the party a pass and immediately returned to the car. Even after Dean had explained to the parking attendant that we had only parked for a few moments, the attendant demanded we pay $15. Dean refused, got out of the car, and ripped the gate off the pivot mechanism, threw it on the sidewalk and got back into the car and drove off. That was not like Dean, but the unfairness of the random death of his father gripped Dean's heart and made him intolerant of unfairness in any form. The healing journey of losing a father to a bear is a ride down a bumpy road.

With Merlyn's death, a chapter of my life had ended. Time heals wounds, and arctic dreams started coming again, not dreams of charging bears but healing stories of pristine wilderness and me paddling my canoe down a wild river. I was cautious to return, with Merlyn no longer alive, no longer there, the North seemed for the first time to be foreboding. Since 1975, whenever I made a northern trip and registered with the Royal Canadian Mounted Police. I always wrote down "Merlyn Carter" as the name to be contacted in the event I did not return on time. I was always at peace with that. On my Hood River trip with my niece, north of the Arctic Circle and ending in Bathhurst Inlet, the RCMP had questioned me about Merlyn being based so far away in Hay River and asked if I wanted to put down someone closer? I replied with utmost confidence, "Just write him down, if anyone can find me it will be Merlyn Carter." I knew now that level of confidence was gone forever.

But was it? There is something about losing a dear friend that also inspires. In many ways, he remains with me. Life comes with no guarantee. Merlyn led a full life. His youngest son had killed his father's killer. His oldest son climbed back into a Twin Otter and flies with his father's spirit on his shoulders. His daughter's eyes and lips hold her father's kindness and likeness as never before. His wife is still bright-eyed, still feisty and returns to Nonacho with renewed focus. He is gone but in many ways Merlyn is still with us.

I was haunted by the reference Merlyn made to the waterfall he had never seen before until that July day in 1997 when he returned our camp with caribou stew and blueberry pie. I decided to return to Nonacho Lake with my twenty-seven-year-old daughter, Lara, to explore the creek and find Merlyn's waterfall. The same Lara that had fallen off the Nonacho Lake dock when she was a toddler. Merlyn jumped into the lake and rescued her. Coming up with a little girl in one hand while still holding his little notebook above water level with his other hand. Everyone else, including me, were still standing around on the dock unaware the child had fallen into the water. Merlyn was a man of action.

From the Carter's camp, we borrowed a fishing boat and, with the canoe hanging five feet over the bow, we motored 45 miles and camped at the beach of the blueberry pie campsite. At this high latitude, decomposition works slowly. Almost 20 years later, charred wood from my campfire in 1997 was evident and there was still a faint outline on the caribou moss from where our tent had been pitched. In the days that followed, Lara and I paddled upstream, stumbled our way around rapids and through blow-downs of black spruce. We crossed three small lakes searching and discovering each time the inlet and the river to the next lake. We lined the canoe up cataracts and pulled the canoe over beaver dams. On one portage, I found a chert knife casually left six thousand years ago on a flat rock much as we might misplace car keys on a park bench today, the difference only being the lapse in time before the next passerby.

Finally, in the twilight of a long arctic day we came to a widening of the creek. Rounding a point, we gasped at the sight of a most exquisite waterfall. The creek had split into three channels. The northernmost and largest fall was an upside-down bowl of tumbling water as white as a freshly painted ceiling; the second fall was a long rolling cascade with water as lively as popping corn, and the third fall was a twisting surge of water that pulsed like the throbbing of a heart.

Three branches of the falls, three ways Merlyn's life and Merlyn's death had touched mine.

His risk taking. He quit school when he knew his future was in the left seat of an airplane. He started a tourist camp before there were tourists. He hired pilots with potential, not always experience. When facing a choice, I learned from Merlyn to jump for it even before I felt ready, and my regrets will not be the times I goofed, but I will regret the opportunities that, through caution, I let slip by. Second-guessing? I keep that quiet.

When in a dilemma he chose kindness. He never stop, people and was always open to helping others reach th without judgment or a discouraging word. In 1992, I plan...ᵤ ᴛo take Lara, that same daughter, but this time only fourteen-years-old on a 28-day arctic canoe trip down the wild, and at that time practically uncharted Snowdrift River, without even an HF radio. Most of my friends and family believed I was crazy. Merlyn dropped us off his Twin Otter MHR at the headwaters of the Snowdrift River, 150 miles northeast of Nonacho. Without a judgmental comment, he just wrote a reminder in his little notebook of my estimated date of arrival at Nonacho Lake, so he would know when to come looking if we didn't show up. I will continue to work toward that type of unconditional acceptance. Not just supporting someone when I agree with what they are doing, but letting them follow their own path and doing what I can to help.

Merlyn almost never flew his airplanes half-full, and he did not lead a half-full life. He was loved, and he loved life because he wasn't worried about an overload; he embraced overloads. He overloaded his life with accomplishments, skills, adventures, and most of all friendships. My take, I have been given but one life, I shall live that life with passion.

Appendix
Merlyn's Aircraft

Cessna 140 1947-1954 This was Merlyn's father's plane. Merlyn likely flew his earliest pilot-in-command hours with this classic taildragger.

Cessna 170B 1952-1960 Merlyn had a special fondness for 170B's. Many pilots believed this plane was underpowered, but on straight skis Merlyn routinely hauled 600 pounds of fish in this 145 hp marvel. He flipped it once while on skis landing on Great Slave Lake when he hit a snowdrift. It was repaired and continued to be used until it was struck by a barge and destroyed on the Hay River.

Mark V Avro Anson Purchased by Carter Fisheries in 1951. Initially flown by legendary bush pilot Stan McMillan. Merlyn logged many hours in the right seat with McMillan. This war bird was the primary fish hauler for Carter Fisheries in the early years. With ice on the wings, Bobby Maclean attempted a takeoff from the West Channel's winter runway, but instead went into the willows. After ten years of hauling slimy fish the plywood structure of the airplane had lost its integrity, "Nails just slid out of the holes." No effort was made to salvage or repair it. Maclean later died while piloting a crop duster in the Caribbean.

Cessna 180 CF-KOW Purchased new in 1958. Ray Baert bought it 1972. KOW crashed and burned shortly thereafter.

Cessna 195 CF-GSF Owned briefly around 1958-59. Merlyn had problems with this "shaking jake." Maybe the shake was due to a worn crankshaft which eventually broke near Fort Nelson, BC causing the propellor to fall off. The float-equipped airplane landed safely on a lake, and was sold shortly thereafter.

199

Cessna 180 CF-LNF Purchased new in 1959. Sold to Leonard Evans, Meadow Lake approximately 1973. I saw NLF in Alaska in 2016. It was still flying, now with USA call letters N846T.

Piper Apache CF-NTE Owned or leased for a few years in the late sixties. Merlyn had no real use for an Apache, more of a family plane, not a business plane.

Beechcraft 18 (Twin Beech) CF-PJD Bought in 1960. This plane was used extensively to haul fish. On paper, it had a similar payload to the Lockheed 10A. In practice, it carried a much lighter load, mainly because the Beechcraft's smaller and tighter landing gear was an impediment to take-offs on the ice. While leased to another company in 1967, this plane was destroyed in a hangar fire.

Cessna Crane Twin engine similar to the Beech 18, but slightly smaller. It is questionable whether Merlyn owned or even leased this plane, but a Crane was reportedly used by Merlyn to haul fish off Hottah Lake around 1961. It could have been borrowed from the McAvoy brothers.

Lockheed 10A CF-HTV Serial # 5 bought in 1961. Also called "The Electra" this type was made famous by Amelia Earhart on her ill-fated around-the-world expedition in 1937. Perhaps Merlyn's favorite plane, it was extensively used to haul fish and freight. Merlyn had talked about saving this plane to fly to airshows during his retirement. Unfortunately, he flipped it on fuel haul to a "dry" lake bed near Birch Lake west of Yellowknife in 1972. Shortly after the crash the plane exploded into flames. The tail section was retrieved in 2013 by Joe McBryan and featured on the Ice Pilots television program.

Cessna 180 CF-OLD Purchased from Bert Courtney (1 year old) in 1963 and sold in 1974.

Helio Courier CF-IYZ Owned around 1962. It was initially purchased for the Pine Point staking rush. Jim McAvoy tore the gear off on a rough landing in 1965. Heavy on the controls and expensive to operate, the plane was sold shortly after it was repaired.

Noorduyn Norseman Mk. IV CF-SAH Owned briefly in the 1960's. SAH on floats became frozen in the ice of the West Channel. One of Merlyn's pilots, familiar with dynamite suggested using the explosive to free the airplane. The first attempt was unsuccessful because presumably of insufficient dynamite. The second attempt blew the plane into smithereens. Enough was salvaged and recovered for the plane to be restored at a Wetaskin, Alberta museum 50 years later.

Noorduyn Norseman Mk. IV CF-FUU Purchased in 1966 from Burt Berry of Uranium City, Saskatchewan to replace CF-SAH. Sold in 1969 after a mishap by Dallas Lay when load shifted and plane stalled on take-off from the float base in Hay River. One of the noisiest commercial airplanes ever built.

de Havilland-DHC-3 Single Otter CF-IOF In 1954 this plane was bought new from de Havilland by the Norwegian Air Force. Upon delivery, it was dropped on the dock in Norway and returned to Canada for repairs. After being repaired it was sold to Imperial Oil and flew over 11,000 hours in northern Alberta and the Northwest Territories. Bought in 1969 from Imperial Oil by Carter Air Service. Sold in 1993 to Randy D'aoust. In 1999, it was converted into a "turbo" Otter.

Cessna 185 CF-QLS Bought in 1972. QLS was the first 185 in the North which was equipped with the 300 hp engine. It also had the first Wipline 3900 floats. In the words of engineer Doug Johnson, "Bigger engine, plus bigger floats, equals bigger opportunity to overload." Nevertheless, Merlyn never liked Buck 85s claiming he could carry the same load in a 180 without the increased fuel consumption, maintenance costs and noise of the 300hp C-185. QLS was sold in 1976 to Buffalo Airways.

McDonald Douglas DC-3 CF-QHY Purchased in 1972 from Canadian Armed Forces for $10,000 was rarely flown by Merlyn. It was sold in 1979, after lengthy leases, to Plummers Lodge.

Cessna 180 CF-TWN (1972 model) This plane was purchased new in 1973 for $20,000, sold to Ray Langille in 1979 for $40,000, bought back by Merlyn in 1986 for $80,000 and sold in 2012 by Carter family to Boyd Warner for $125,000.

Two supporting struts inside the cabin of TWN were perfect handholds should a passenger and pilot decide to join "the mile-high club."

de Havilland DHC-3 Single Otter C-FCZP Purchased in 1973 from Northward Aviation. Sold in 1984 to Raecom Air. Equipped with floats and wheel/skis. de Havilland only produced 450 Single Otters (1951-1967) mostly for the military market. Many are still flying worldwide.

de Havilland DHC-6 series 100 Twin Otter CF-MHR Purchased in 1978. Sold in 1990 to Landa Aviation. Equipped with floats and wheel/skis

de Havilland DHC-6 series 100 Twin Otter CG-KAZ Purchased in 1984 from Ptarmigan Airways. Sold in 1996 to Peruvian military.

Cessna 140 CF-YNC Purchased in 1984. Sold in 1991. Sold shortly after carb ice or ice in the fuel line caused Merlyn to make an emergency landing on Great Slave Lake 14 miles from Fort Resolution.

Cessna Caravan 208 CG-JEM This plane was equipped with amphibious floats to facilitate water and runway landings. Purchased in 1997 in Perth, Australia. Sold to Doug Williamson in 2002.

Other Aircraft mentioned in Book

Cessna 206 CG-VCM The 206 has replaced the Cessna 185 for most small plane float applications. It's larger cabin and modernized design is a positive. However, it's tricycle gear makes it a much less desirable bush plane when on wheels. However, the 206 equipped with big squishy tundra tires is commonly used by Alaskan bush pilots, who frequently land on gravel bars. Eric Sieh is making a name for himself flying a 206 on tundra tires out of Kotzebue, AK.

de Havilland DHC Beaver STOL bush plane with a fabled reputation.

McDonald Douglas DC-6 Merlyn had the plan to buy a 4-engine DC-6 with zero-time engines. These obsolete planes, when the engines are timed out, could be bought at an auction for next to nothing. Merlyn figured he could get one last flight out of one. Merlyn planned to fill it with barrels of fuel for the planes, outboard motors and camp powerplant, 7,500 gallons of fuel would have kept the camp supplied for several seasons. He would fly it to Nonacho Lake in spring, land on the ice and taxi it up on the beach and leave it there, as a fuel cache and landmark. There were a few bugs in this plan that he was still working out at the time of his death.

Piper PA-20 Pacer CF-XXO An early 1950s fabric precursor of the famed Piper Tri-Pacer, the Pacer was short coupled 4-place, taildragger that was prone to ground-looping on wheels. When equipped with 1650 floats it was prone to sinking a float when taxiing downwind. The author of this book owned a Pacer for several years while he lived in Fort Resolution and Hay River and based the plane at Carter's Nonacho Lake camp one summer.

Piper J3 Cub The classic fabric, tandem, taildragger trainer, but too small for almost all commercial applications.

Stinson Most pilots will shake their head the first time they see a gull wing Stinson, wondering how that fat wing was ever designed, and if it could actually fly! But it did, and Stinsons were quality built, but seldom seen airplanes in the North.

Fairchild Although Merlyn never owned one, Fairchilds were common commercial bush planes in the early years.

(photo courtesy of Kandee Froese)

Acknowledgements

I met Merlyn Carter in 1975. The summer before, my first wife, Bonnie, and I had paddled over 1,000 miles on the Liard and Mackenzie Rivers. We had come to Hay River to live a dream. Our plan was to sell our Toyota Land Cruiser and our Old Town Tripper canoe. With the proceeds, we could buy a 20-foot freighter canoe, an outboard motor, and head down the Mackenzie again and up the Great Bear River to Great Bear Lake. We planned to build a log cabin and spend a year on the wilderness shore of Great Bear Lake, but we hit a snag. Our vehicle and canoe were registered in Colorado, and it was not legal or practical to sell them to Canadian citizens. We had no money, no future and sat dejected at the end of the road in Hay River. We sat with our feet dangling over the breakwater that was the dock for Carter Air Service's four floatplanes. Merlyn walked up to us as he was making notes in a tiny black book. He stuffed the little book in his shirt pocket and patiently listened to our plight.

"Why do you want to go up to Great Bear? Some summers the ice does not even completely come off that lake." We explained that we were intrigued by the crystal-clear water of the Great Bear River and figured Great Bear Lake would be a pristine spot for a cabin. Immediately Merlyn said, "I've got a camp on Nonacho Lake, same blue water, same rock, and more sand, very nice out there." He paused for just a second and went on to offer up a suggestion. "I'll tell you what. I have lots of loads going out to my camp, but sometimes just a half-load. You can stick around here if you want, and when I get a half-load, I'll take you two and your little canoe out there, and you can paddle around Nonacho and find a place to build your cabin."

We were ecstatic, actually jumping in joy. He went on, "While you are waiting, you two can sleep in that little green shed over there and help me load and unload the planes, pump floats and fuel up my planes, and maybe next winter out at the camp you can help me fill the icehouse."

207

Two 23-three-year-olds could not have said "yes" faster. We moved into the shed. The plywood floor was stained by avgas and motor oil. Our sleeping bags were shadowed beneath the Otter's wheel-skis which were stored awaiting the winter landing gear switchover. It was perfect. We became the young American couple that lived in the shed, and we befriended Merlyn's kids and pilots.

We crawled on top of the Cessnas' wings dragging heavy rubber fuel hoses and filling the tanks. The Otters had belly tanks and we got soaked in avgas a few times when the fuel sloshed back out the hole until we learned to do it right.

For two-weeks we scrambled around on floats removing rubber balls that had been cut in half to seal the float holes. Next, with a little gray plastic hand pump, we would pump out any water that had accumulated from leaky float rivets or from the flexing of the aluminum seams on take-off and landing. Merlyn would always say, "A pound of water in the floats is like two pounds in the cabin."

We met the planes as they came in, unloaded tubs of fish caught by commercial fishermen and loaded them into the bed of Merlyn's pick-up truck. Merlyn would just throw me his truck keys, and I would drive the pick-up to the Freshwater Fish plant in the new town and deliver the fish.

More often, the airplanes' loads were coming in from and going out to Carter's sport fishing camps. Duffels of gear, sport fishing enthusiasts, their grub and beer going in and happy faces and big trophy fish coming out. But every time it looked like there might be space for us going in to the lake, Merlyn always needed more fuel for the camp. Instead of us, we would roll big drums of pink 80/87 avgas into the Single Otters until the floats were almost underwater. Then Merlyn would climb in and with a backfire and puff of smoke the big radial engine would start to thumpity thump, and we would push on the struts as the seaplane left us behind.

On July 5th, we had enough. Discouraged, we moved out of the green shed, loaded up our Land Cruiser and without a word to Merlyn headed south. We only got as far as the caboose at the edge of town and turned around. Bonnie was in tears. We would give it one more try. I don't know if Merlyn sensed our frustration and impatience, but two days later he flew us and our entire outfit to Nonacho Lake.

We paddled around a small portion of the lake and decided on a sandy hilltop above a beach for our cabin site. Not by intention but by happenstance it was only a mile from Carter's place. Choosing a homesite in close proximity of the Carter camp was the best decision I have ever made.

Although, we spent most of the summer building our log cabin we did get over to Carters at least once a week. Merlyn's wife Jean came out with the family in September to pick cranberries, and we became close friends with all the Carters.

Life has its ups and downs, and through it all, for 30 years, Merlyn was my best friend. Merlyn and Jean saved my oldest daughter's life when she was just an infant. You never forget something like that. I wanted to write this book for a long time. I wish I had written it while Merlyn was still alive. But better late than never.

When I first proposed the idea to a publisher, he said, "Biographies don't sell."

I implored him to listen to Merlyn's story, 25,000 hours of bush flying, the adventures, the mishaps, the fatal ending at the claws of a bear.

He said, "It takes three things for a book to sell. #1 You've got to have a good story. #2 You've got to be good looking. #3 You've got to have a platform to sell from." He went on to say, "And I did not necessarily state them in order of importance. All you have is a good story."

I wrote the book anyway. I had to. It is a good story and it is a story that needs to be told. I put my heart into this book, but it would not have been possible without the support, input, and cooperation of many.

For seed money, I put out a request on the crowd sourcing website Kickstarter. The response was overwhelming, and the generosity was beyond my wildest expectations. The support enabled me to travel to Saskatchewan and the Northwest Territories to interview Merlyn's family and acquaintances. More than the expense money, not completing the "The Merlyn Project" would have let down all my backers. I couldn't live with the guilt. So I persevered. Completing the manuscript was more difficult and time consuming than I could have possibly imagined. My partner, Sue Plankis has infinite patience, and she needed it when I was stuck in front of the computer screen trying and trying to put this story together. All those Kickstarter supporters were counting on me and although they were always supportive of my updates I know I must have tested their patience.

This project would never have happened without the generous support of my Kickstarter backers. Thank you for believing in me: (in alphabetic order) Christopher Alan, Mike Bartz, Dick Bellman, Mary Jean Blaisdell, Ivan Bourque, Mike Branham, Jon Breimhorst, Anne and Larry Buckmaster, David Burkhart, Myles Carter, Gary Cobus, Dan Cooke, Karen Cooke, Jan Curtice, Chuck Davidge, Mark Deering, Jan Dettmer, Keith Dohey, Draper, Chris Dzombak, Eileen Fisher, M. Fisher, Jim Fitzpatrick, Beth Flaherty, G. Darcy Fleming, Ryan Fox, Sarah Froese, Christopher Gaul, Paul Grondin, Carmen Hannon, Deb Harley, Laura Held-Honeck, Penni Schacher Hess, Kathy Holladay, Tina Howard, Patrick Hueller, Cliff Jacobson, Damon Johnson, Karen Kelley, Lori Kelley, Rich Kelley, Clark Kent, Mari Kesselring, Trinity Kesselring, Zoe Kesselring, Timothy Koestler, Tom Koshiol, John Kuhfahl, David Lasslett, Luke, Nancy Makepeace, Deb and John Markham, Valerie Miller Markham, Mary McCordic, Scott McCready, Mark McGuire, Kate Merriweather, Mark Morrall, Riajan Musselman, Will Nagle, Bob O'Hara, Darlene Patterson, Sue Plankis, Stu Poirier, Brenda Postels, Piiskoor Prof, Dick Pula, L.E. Quenemoen, T. Rauscher, Allan and Diane Reid, Michael B. Roy, Dan Schwartz, Tim Schwolert, Fred Shermock, Mike Smith, Chip Sutton, Diane Temple, Susan Temple, Greg Towe, James Van Patten, Linda Wallenberg, Eric Damon Walters, Henry T. Wang, Ellen Winchester, Gray Wolf, Sarah Woodcock, Pam Wright.

I also want to thank those that have contributed to the content of this book. I will shout out a few in particular, but all of you in some way helped my dream take shape: (in alphabetic order)

Kevin Arychuck, Tim Beahen, Richard Berg, Pat Bobinski. Bert Buckley. Anne Buckmaster, Larry Buckmaster, Dean Carter, Gary Carter, Jean Carter, Myles Carter, Roy Carter, Dave Cathcart, Gordon Clark, Chuck Davidge, Ed Eckstrom, Jason Froese, Kandee Froese, Randy Froese, Colleen Gagnier, Joe Gagnier, Michael Gagnier, Clay Gamble, Andy Greenblatt, Jim Harrison, Cliff Jacobson, Tony Jarrod, Andy Jenks, Doug Johnson, Kail Katzenmeier, Karen Kelley, Pam Kelley, Mari Kesselring, Julie Kilpatrick, Norman and Elsie Lay, Alex Leliberte, Perry Linton, Phylis Linton, Ella Mae Maher, Jim Maher, Tom Makepeace, Deb Markham, Celine Marlow, George Marlow, Joe McBryan, Diane McCallum, Marius McCallum, Alec Morin, Bob O'Hara, Charlie Perkins, Sue Plankis, Stewart Poirier, Ron Rogers, Alex Schamber and the Goodsoil Historical Museum, Elaine Sergent, Samuel Wosmek, Reg Workum.

Jean Carter welcomed me to her home on Vale Island in Hay River, and we spent a week together looking through photos and documents and recording her memories. Myles Carter, Kandee Froese, and Dean Carter were also instrumental in getting the stories, dates, and photos straight.

Marius McCallum took me out commercial fishing in December on Great Slave Lake. Despite the calendar reading 2015 the technique and technology had barely changed since the 1950s. The bug we travelled in was built in the 1950s, the jigger, the nets, and the tools had changed little. Driving a bug in pre-dawn darkness on a day with the temperature below zero, pulling fish out of the nets, the blood, the guts the sweat it helped me understand the business. I could only wish Merlyn would have banked in tight and landed with the old Anson and picked up a load of fish.

I met with Joe McBryan (Buffalo Joe from the hit television program Ice Pilots) after he climbed out of the cockpit of one of his DC-3's. Although he said he only had a moment he talked with me for two hours. McBryan talked with clarity about the early days of hauling fish when he worked and flew with Merlyn.

Norman and Elsie Lay welcomed me into their home and spent an afternoon sharing the Lay homesteading experience and took me on a tour of the old homestead and family graveyard. To view the Sections that were once forested wilderness, but now transformed into fertile fields of grain was amazing.

Accomplished pilots of a newer era, Larry Buckmaster and Dean Carter helped with some of the technical aspects of Merlyn's aircraft.

I thank Susan Temple for coming with me to Hay River for Merlyn's memorial service and for her unconditional support and her understanding during the months following Merlyn's passing.

Karen Kelley, Cliff Jacobson, Deb Markham, Pam Kelley all helped with editing and supported the concept of the book.

Andy Jenks was a stalwart helper with tech issues and especially helpful in creating the custom maps of Merlyn's flying territory. His uncle, Samuel Wosmek, did the Single Otter artwork that is used to tie the story together. I appreciate their generous and talented contributions.

CPSIA information can be obtained
at www.ICGtesting.com
Printed in the USA
BVHW04s1751200818
524806BV00005B/6/P